SpringerBriefs in Electrical and Computer Engineering

Series editors

Woon-Seng Gan, School of Electrical and Electronic Engineering, Nanyang Technological University, Singapore, Singapore
C.-C. Jay Kuo, University of Southern California, Los Angeles, CA, USA
Thomas Fang Zheng, Research Institute of Information Technology, Tsinghua University, Beijing, China
Mauro Barni, Department of Information Engineering and Mathematics, University of Siena, Siena, Italy

SpringerBriefs present concise summaries of cutting-edge research and practical applications across a wide spectrum of fields. Featuring compact volumes of 50 to 125 pages, the series covers a range of content from professional to academic. Typical topics might include: timely report of state-of-the art analytical techniques, a bridge between new research results, as published in journal articles, and a contextual literature review, a snapshot of a hot or emerging topic, an in-depth case study or clinical example and a presentation of core concepts that students must understand in order to make independent contributions.

More information about this series at http://www.springer.com/series/10059

Ahmed Masmoudi

Control Oriented Modelling
of AC Electric Machines

 Springer

Ahmed Masmoudi
Sfax Engineering National School
University of Sfax
Sfax
Tunisia

ISSN 2191-8112   ISSN 2191-8120   (electronic)
SpringerBriefs in Electrical and Computer Engineering
ISBN 978-981-10-9055-4   ISBN 978-981-10-9056-1   (eBook)
https://doi.org/10.1007/978-981-10-9056-1

Library of Congress Control Number: 2018936658

Printed on acid-free paper

This Springer imprint is published by the registered company Springer Nature Singapore Pte Ltd. part of Springer Nature
The registered company address is: 152 Beach Road, #21-01/04 Gateway East, Singapore 189721, Singapore

Preface

Following the 70th oil crisis, the world realized for the first time what it would be like if fuels would no longer be cheap or unavailable. In order to damp the fallouts of such a situation, renewable energies have been the subject of an intensive regain of interest. So many R&D programmes were launched so far, with emphasis on the investigation of the power potential of conventional (wind, solar and biomass) and emergent (marine, geothermal) earth's natural energy reserves.

Moreover, until the 1960s, automotive manufacturers did not worry about the cost of fuel. They had never heard of air pollution, and they never thought about life cycle. Ease of operation with reduced maintenance costs meant everything back then. In recent years, clean air policies are driving the market to embrace new propulsion systems in an attempt to substitute or to assist efficiently the internal combustion engine (ICE) by an electric drive unit, yielding respectively the so-called electric and hybrid propulsion systems.

The above sustainable energy and mobility applications consider in most if not all cases a key component that achieves the electro-mechanical conversion of energy: the electric machine. It operates as a generator which converts directly converts the wind and wave energies, and through a turbine the solar, biomass and geothermal ones, into electricity. It operates as a propeller fed by a battery or a fuel cell pack embedded on board of electric and hybrid vehicles.

This said, it should be underlined that the machine integration in the above-cited and the overwhelming majority of current applications represents a symbiosis of several engineering fields with a dominance of the electrical one. Of particular interest are machine control strategies thanks to which variable speed drives and generators are continuously reaching higher and higher degrees of performance. This has been systematically initiated by the selection of appropriate and accurate models of the machines to be controlled.

Within this trendy topic, the manuscript deals with the modelling of AC machines, in so far as they are currently equipping the major part of the variable speed drives and generator; the dc machines are doomed to disappear in a near future. The manuscript is structured in two chapters:

- The first one is aimed at modelling of the induction machine considering its a-b-c and *Park* models. An analysis of the machine steady-state operation is then carried out using its *Park* model. A case study dealing with the doubly fed induction machine, a viable candidate for wind power generating systems, is treated with emphasis on a typical topology in which the brush-ring system is discarded, yielding the so-called brushless cascaded doubly fed machines.
- The second chapter is devoted to the modelling of the synchronous machines, with emphasis on its a-b-c and *Park* models. A special attention is paid to the formulation and analysis of the electromagnetic torque with an investigation of the variations of its synchronizing and reluctant components in terms of the torque angle. The chapter is achieved by a case study dealing with an investigation of the main features of the electric drive unit of a hybrid propulsion system and the possibility of extending the flux weakening range of the propeller which is made up of the PM synchronous motor.

Sfax, Tunisia   Prof. Ahmed Masmoudi
                                Head of the Renewable Energies and Electric Vehicles Lab

Contents

About the Author

Ahmed Masmoudi received his B.S. degree from Sfax Engineering National School (SENS), University of Sfax, Sfax, Tunisia, in 1984; Ph.D. degree from Pierre and Marie Curie University, Paris, France, in 1994; and the Research Management Ability degree from SENS, in 2001, all in electrical engineering. In August 1984, he joined Schlumberger as a field engineer. After this industrial experience, he joined the Tunisian University where he held different positions involved in both education and research activities. He is currently a Professor of electric power engineering at SENS, the Head of the Research Laboratory on Renewable Energies and Electric Vehicles (RELEV) and the Coordinator of the master on Sustainable Mobility Actuators: Research and Technology. He published up to 85 journal papers, 19 among which in IEEE transactions. He presented up 367 papers at international conferences, 9 among which in plenary sessions and 3 have been rewarded by the best presented paper prize. He is the co-inventor of a USA patent. He is the Chairman of the Programme and Publication Committees of the International Conference on Ecological Vehicles and Renewable Energies (EVER), organized every year in Monte Carlo, Monaco, since 2006. He was also the Chairman of the Technical Programme and Publication Committees of the First International Conference on Sustainable Mobility Applications, Renewables, and Technology (SMART) which has been held in Kuwait in November 2015. Its involvement in the above conferences has been marked by an intensive guest-editorship activity with the publication of many special issues of several journals including the IEEE Transactions on Magnetics, COMPEL, ELECTROMOTION and ETEP. He is a senior member of IEEE. His main interests include the design of new topologies of AC machines allied to the implementation of advanced and efficient control strategies in drives and generators, applied to renewable energy as well as to electrical automotive systems.

List of Figures

Chapter 1
Induction Machine Modelling

Abstract The chapter deals with the modelling of the induction machine (IM). Following the analysis of the principle of operation which is based on the induction phenomenon, the a-b-c model is established assuming a sinusoidal spatial repartition of the air gap flux density, a linear magnetic circuit, and constant phase resistors. The a-b-c model makes possible the establishment of a state representation of the IM. Then, the Park transform is introduced and applied to the IM a-b-c model, leading to its Park one. An analysis of the IM steady-state operation is then carried out using its Park model. The chapter is achieved by a case study dealing with the doubly fed induction machine which is widely integrated in wind power generating systems.

Keywords Induction machine · Modelling · A-B-C model · State representation
Park Model

1.1 Introduction

Generally speaking, the modelling of a system is an approach to formulate its behaviour by a set of equations. The model allows, for a given domain of validity and a given accuracy, the prediction of the system outputs in terms of its inputs. The modelling of electric machines has been and remains a state-of-the-art topic. It represents a crucial step to initiate any R&D project aimed at the design or the control or both, of electric machines.

Concerning the machine design topic, the modelling is achieved by means of:

- Numerical approaches they are aimed at the resolution of the *Maxwell*s' equations considering different numerical procedures. The most popular one is the finite element method (FEM), also named finite element analysis (FEA). FEA-based design of electric machines is reputed by its high accuracy. However, it is reserved to specialists and is a great consumer of CPU time,

- Analytical approaches they consider different formulations involving the machine geometry and materials, taking into consideration well-known electromagnetic laws. Of particular interest is the *Hopkinson* law on the basis of which is developed

© The Author(s) 2018   1
A. Masmoudi, *Control Oriented Modelling of AC Electric Machines*,
SpringerBriefs in Electrical and Computer Engineering,
https://doi.org/10.1007/978-981-10-9056-1_1

the most popular analytical modelling approach, that is the magnetic equivalent circuit (MEC), also called lumped circuit. The MEC modelling is a powerful tool for the machine preliminary design and sizing. It is rapid and leads to acceptable accuracy,
- Combined analytical–numerical approaches In order to improve the accuracy of the MEC models, some features whose analytical prediction is critical are offline computed by FEA and are provided to the MEC-solver. These approaches represent the best compromise rapidity/accuracy.

Regarding the machine control topic, the modelling is mostly done analytically. However and accounting for the high nonlinearities involved in the machine models, their resolution is achieved numerically considering different algorithms, such as the *Runge-Kutta* one. Most if not all machine models include (i) an electrical equation, (ii) a magnetic equation, and (iii) a mechanical equation. These involve the machine phase variables, leading to the so-called a-b-c models. In order to simplify the synthesis of the control laws, mathematical transformations are commonly applied. These enable to substitution of the a-b-c models by two-phase ones. The most popular transformation is the one introduced by *Park*.

The chapter deals with the modelling of the induction machine (IM) considering its a-b-c and *Park* models.

1.2 Principle of Operation: Induction Phenomenon

Let us consider the case of a wound rotor three-phase induction machine, with:
- its stator circuits fed by three-phase balanced currents with an angular frequency ω_s,
- its rotor circuits are kept open.

Doing so, a rotating field takes place in the air gap that has the speeds:
- Ω_s with respect to the stator, with $\Omega_s = \frac{\omega_s}{p}$ where p is the IM pole pairs,
- $\Omega_{s/r}$ with respect to the rotor.

Giving the fact that the rotor circuits are open, the torque production condition is not fulfilled and the shaft remains stationary, leading to:

$$\Omega_{s/r} \; = \; \Omega_s = \frac{\omega_s}{p} \tag{1.1}$$

The rotating field induces three back-EMFs in the rotor circuits that have an angular frequency ω_r, with:

$$\omega_r = P\Omega_{s/r} \; = \; \omega_s \tag{1.2}$$

The similarity of the angular frequencies of both stator and rotor circuits yields the so-called transformer operation of the IM. The motor operation is accessed when the torque production condition is met. For that, the rotor circuits have to be closed/short-circuited, resulting in three-phase balanced currents with an angular frequency ω_r. Hence, a second rotating field takes place in the air gap that has the speeds:

- $\Omega_{r/s}$ with respect to the stator,
- Ω_r with respect to the rotor, with $\Omega_r = \frac{\omega_r}{p} = \frac{\omega_s}{p}$

As a summary, the air gap is traversed by two fields rotating synchronously at the speed $\frac{\omega_s}{p}$ with respect to the rotor. Consequently, an electromagnetic torque is produced and the shaft starts rotating in the direction of the rotating field resulting from the interference between the rotor and stator rotating fields. Following the start-up, a steady state is reached, characterized by a rotor speed Ω_m. Thus, the stator rotating field speed $\Omega_{s/r}$ turns to be:

$$\Omega_{s/r} = \Omega_s - \Omega_m \tag{1.3}$$

The angular frequency of the back-EMFs induced in the rotor circuits is then:

$$\omega_r = p(\Omega_s - \Omega_m) = \omega_s - p\Omega_m \tag{1.4}$$

The speed of the rotating field created by the rotor, with respect to the rotor, becomes:

$$\Omega_r = \frac{\omega_r}{p} = \Omega_s - \Omega_m \tag{1.5}$$

Thus, the synchronism between the two rotating fields is kept and the torque production condition is met for all rotor speeds, except for $\Omega_m = \Omega_s$ for which the induction phenomenon disappears. Indeed, there are no back-EMFs induced in the rotor circuits, then no rotor currents and no rotating field created by the rotor circuits, and consequently the torque production condition is no longer fulfilled.

In order to characterize the steady-state shift between the speed of the resultant rotating field with respect to the stator Ω_s and the rotor one Ω_m, a slip s is commonly considered, such that:

$$s = \frac{\Omega_s - \Omega_m}{\Omega_s} = \frac{\omega_r}{\omega_s} \tag{1.6}$$

1.3 Model Simplification Hypothesis

AC machine models have been commonly simplified considering given hypothesis, especially:

- The MMFs created within the air gap, by the different circuits (single or multi-phase), are assumed to have sinusoidal spatial repartitions. Thus, the effects of the

spatial harmonics on the developed electromagnetic torque or on the generated back-EMFs are omitted,
- The magnetic circuit is supposed linear (unsaturated). That is to say that the fluxes created by the different circuits remain proportional to the currents that generated these fluxes. Consequently, the self and mutual inductances characterizing these circuits turn to be independent of their currents,
- The resistances of the different circuits are assumed constant by neglecting the heating and skin effects. It should be noted that the circuits installed inside the machine are subject of heating due to *Joules* and iron losses, leading to an increase of their resistances. The *Joules* losses increase with the load. The iron losses increase with the frequency, so does the skip effect. Indeed, under high frequencies, the electrons tend to circulate at the skin of the conductors rather than in their whole section at low frequency. Thus, the resistances increase with the skin effect,
- In smooth pole machines (induction and smooth pole synchronous machines), the slotting effect is neglected assuming a constant air gap. Consequently, the winding self and mutual (between its circuits) inductances are independent of the rotor position.

1.4 IM A-B-C Model

The a-b-c model considers the electrical variables applied to and measured in the IM circuits. These variables are depicted in Fig. 1.1 which gives a schematic representation of a three-phase IM.

1.4.1 Electrical Equation

Referring to Fig. 1.1, the application of the *Ohm* law gives the following equation:

$$V = RI + \frac{d}{dt}\Phi \qquad (1.7)$$

where V, I, et Φ are the voltage, current, and flux vectors, respectively. These could be expressed as follows:

$$V = \begin{bmatrix} V_s \\ V_r \end{bmatrix} \qquad I = \begin{bmatrix} I_s \\ I_r \end{bmatrix} \qquad \Phi = \begin{bmatrix} \Phi_s \\ \Phi_r \end{bmatrix} \qquad (1.8)$$

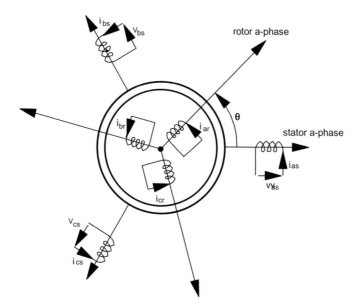

Fig. 1.1 Schematic representation of the induction machine

with:

$$\begin{cases} V_s = \begin{bmatrix} v_{as} \\ v_{bs} \\ v_{cs} \end{bmatrix} & I_s = \begin{bmatrix} i_{as} \\ i_{bs} \\ i_{cs} \end{bmatrix} & \Phi_s = \begin{bmatrix} \phi_{as} \\ \phi_{bs} \\ \phi_{cs} \end{bmatrix} \\ V_r = \begin{bmatrix} 0 \\ 0 \\ 0 \end{bmatrix} & I_r = \begin{bmatrix} i_{ar} \\ i_{br} \\ i_{cr} \end{bmatrix} & \Phi_r = \begin{bmatrix} \phi_{ar} \\ \phi_{br} \\ \phi_{cr} \end{bmatrix} \end{cases} \tag{1.9}$$

and where R is the resistance matrix, such that:

$$R = \begin{bmatrix} r_s \mathcal{I}_3 & \mathcal{O}_3 \\ \mathcal{O}_3 & r_r \mathcal{I}_3 \end{bmatrix} \tag{1.10}$$

where:

- r_s et r_r are the stator and rotor phase resistances, respectively,
- $\mathcal{I}_3$ and $\mathcal{O}_3$ are the identity and the null matrices of rank 3, respectively.

1.4.2   Magnetic Equation

In Eq. (1.7), the flux and current vectors are linked by the following expression:

$$\Phi = LI \tag{1.11}$$

where L is the inductance matrix, such that:

$$L = \begin{bmatrix} L_{ss} & L_{sr} \\ L_{rs} & L_{rr} \end{bmatrix} \tag{1.12}$$

with:

$$L_{ss} = \begin{bmatrix} L_s & M_s & M_s \\ M_s & L_s & M_s \\ M_s & M_s & L_s \end{bmatrix} \qquad L_{rr} = \begin{bmatrix} L_r & M_r & M_r \\ M_r & L_r & M_r \\ M_r & M_r & L_r \end{bmatrix} \tag{1.13}$$

and:

$$L_{sr} = L_{rs}^t = M_{rs} \begin{bmatrix} \cos\theta & \cos\left(\theta + \frac{2\pi}{3}\right) & \cos\left(\theta - \frac{2\pi}{3}\right) \\ \cos\left(\theta - \frac{2\pi}{3}\right) & \cos\theta & \cos\left(\theta + \frac{2\pi}{3}\right) \\ \cos\left(\theta + \frac{2\pi}{3}\right) & \cos\left(\theta - \frac{2\pi}{3}\right) & \cos\theta \end{bmatrix} \tag{1.14}$$

where:
$\begin{cases} \theta & : \text{the electrical angular displacement of the rotor with respect to the stator,} \\ L_s & : \text{the stator phase self-inductance,} \\ L_r & : \text{the rotor phase self-inductance,} \\ M_s & : \text{the mutual inductance between two stator phases,} \\ M_r & : \text{the mutual inductance between two rotor phases,} \\ M_{rs} & : \text{the maximum value of the mutual inductance between a stator phase and} \\ & \quad \text{a rotor one which is reached when their magnetic axis are aligned.} \end{cases}$

Accounting for relation (1.11), the electrical Eq. (1.7) could be rewritten as follows:

$$\begin{bmatrix} V_s \\ V_r \end{bmatrix} = \begin{bmatrix} r_s \mathcal{I}_3 & \mathcal{O}_3 \\ \mathcal{O}_3 & r_r \mathcal{I}_3 \end{bmatrix} \begin{bmatrix} I_s \\ I_r \end{bmatrix} + \frac{d}{dt} \left\{ \begin{bmatrix} L_{ss} & L_{sr} \\ L_{rs} & L_{rr} \end{bmatrix} \begin{bmatrix} I_s \\ I_r \end{bmatrix} \right\} \tag{1.15}$$

1.4.3   Mechanical Equation

The mechanical equation is derived from the dynamic fundamental principle, as:

$$T_{em} - T_l = J\frac{d\Omega_m}{dt} \tag{1.16}$$

where:
$$\begin{cases} \Omega_m & : \text{the rotor mechanical speed,} \\ T_{em} & : \text{the electromagnetic torque,} \\ T_l & : \text{the load torque,} \\ J & : \text{the moment of inertia.} \end{cases}$$

The rotor speed Ω_m is expressed in terms of θ, as follows:

$$\Omega_m = \frac{1}{p}\frac{d\theta}{dt} \tag{1.17}$$

The electromagnetic torque has been formulated in [1], as follows:

$$T_{em} = \frac{1}{2} p\, I^t \left\{ \frac{d}{d\theta}[L] \right\} I \tag{1.18}$$

which could be developed as:

$$T_{em} = \frac{1}{2} p \begin{bmatrix} I_s \\ I_r \end{bmatrix}^t \begin{bmatrix} 0 & \frac{d}{d\theta}[L_{sr}] \\ \frac{d}{d\theta}[L_{rs}] & 0 \end{bmatrix} \begin{bmatrix} I_s \\ I_r \end{bmatrix} \tag{1.19}$$

Knowing that:

$$I_s^t \left\{ \frac{d}{d\theta}[L_{sr}] \right\} I_r = I_r^t \left\{ \frac{d}{d\theta}[L_{rs}] \right\} I_s \tag{1.20}$$

the expression of the electromagnetic torque turns to be:

$$C_{em} = p\, I_r^t \left\{ \frac{d}{d\theta}[L_{rs}] \right\} I_s \tag{1.21}$$

Assuming that the stator and rotor currents are balanced, the following relations could be expressed:

$$\begin{cases} i_{as} + i_{bs} + i_{cs} = 0 \\ \\ i_{ar} + i_{br} + i_{cr} = 0 \end{cases} \tag{1.22}$$

The development of Eq. (1.21), taking into account the relations given in (1.22), leads to the following expression [2]:

$$T_{em} = 3 \, P \, M_{rs} \left(i_{as} i_{br} \sin(\theta - \frac{2\pi}{3}) + i_{bs} i_{ar} \sin(\theta + \frac{2\pi}{3}) - (i_{as} i_{ar} + i_{bs} i_{br}) \sin \theta \right) \tag{1.23}$$

1.4.4   State Representation

The electrical Eq. (1.7) is rewritten taking into account the current-flux relation, as:

$$V = RL^{-1}\Phi + \frac{d}{dt}\Phi \tag{1.24}$$

Hence, the association of Eqs. (1.24) and (1.16) leads to a state representation where the flux, the electrical position, and its derivative with respect to time are the state variables. The development of Eq. (1.24) requires the determination of the inverse of the inductance matrix which will be carried out in what follows.

Assuming balanced fluxes in both stator and rotor phases, one can establish the following:

$$\begin{cases} \phi_{as} + \phi_{bs} + \phi_{cs} = 0 \\ \\ \phi_{ar} + \phi_{br} + \phi_{cr} = 0 \end{cases} \tag{1.25}$$

Accounting for relations (1.22) et (1.25), the current-flux one (1.11) is reduced to:

$$\Phi_1 = L_1 I_1 \tag{1.26}$$

where:

$$\Phi_1 = \begin{bmatrix} \Phi_{1s} \\ \Phi_{1r} \end{bmatrix} \qquad\qquad I_1 = \begin{bmatrix} I_{1s} \\ I_{1r} \end{bmatrix} \tag{1.27}$$

with:

$$\Phi_{1s} = \begin{bmatrix} \phi_{as} \\ \phi_{bs} \end{bmatrix} \quad I_{1s} = \begin{bmatrix} i_{as} \\ i_{bs} \end{bmatrix} \quad \Phi_{1r} = \begin{bmatrix} \phi_{ar} \\ \phi_{br} \end{bmatrix} \quad I_{1r} = \begin{bmatrix} i_{ar} \\ i_{br} \end{bmatrix} \tag{1.28}$$

and where:

$$L_1 = \begin{bmatrix} L_{ss1} & L_{sr1} \\ L_{rs1} & L_{rr1} \end{bmatrix} \tag{1.29}$$

with:

$$\begin{cases} L_{ss1} = \begin{bmatrix} L_s - M_s & 0 \\ 0 & L_s - M_s \end{bmatrix} \\[2mm] L_{rr1} = \begin{bmatrix} L_r - M_r & 0 \\ 0 & L_r - M_r \end{bmatrix} \\[2mm] L_{sr1} = M_{rs} \begin{bmatrix} \cos\theta - \cos(\theta - \frac{2\pi}{3}) & \cos(\theta + \frac{2\pi}{3}) - \cos(\theta - \frac{2\pi}{3}) \\ \cos(\theta - \frac{2\pi}{3}) - \cos(\theta + \frac{2\pi}{3}) & \cos\theta - \cos(\theta + \frac{2\pi}{3}) \end{bmatrix} \\[4mm] L_{rs1} = M_{rs} \begin{bmatrix} \cos\theta - \cos(\theta + \frac{2\pi}{3}) & \cos(\theta - \frac{2\pi}{3}) - \cos(\theta + \frac{2\pi}{3}) \\ \cos(\theta + \frac{2\pi}{3}) - \cos(\theta - \frac{2\pi}{3}) & \cos\theta - \cos(\theta - \frac{2\pi}{3}) \end{bmatrix} \end{cases} \tag{1.30}$$

Let us call $l_s = L_s - M_s$ and $l_r = L_r - M_r$. The development of matrix L_1 gives:

$$L_1 = \sqrt{3}M_{rs} \begin{bmatrix} \frac{l_s}{\sqrt{3}M_{rs}} & 0 & \sin(\theta + \frac{2\pi}{3}) - \sin\theta \\ 0 & \frac{l_s}{\sqrt{3}M_{rs}} & \sin\theta & -\sin(\theta - \frac{2\pi}{3}) \\ -\sin(\theta - \frac{2\pi}{3}) & \sin\theta & \frac{l_r}{\sqrt{3}M_{rs}} & 0 \\ -\sin\theta & \sin(\theta + \frac{2\pi}{3}) & 0 & \frac{l_r}{\sqrt{3}M_{rs}} \end{bmatrix} \tag{1.31}$$

which could be rewritten as:

$$L_1 = \begin{bmatrix} l_s \mathcal{I}_2 & \sqrt{3}M_{rs} A \\ \sqrt{3}M_{rs} B & l_r \mathcal{I}_2 \end{bmatrix} \tag{1.32}$$

where $\mathcal{I}_2$ is the identity matrix of rank 2 and where matrixes A and B are defined as follows:

$$\begin{cases} A = \begin{bmatrix} \sin(\theta + \frac{2\pi}{3}) & -\sin\theta \\ \\ \sin\theta & -\sin(\theta - \frac{2\pi}{3}) \end{bmatrix} \\ \\ B = \begin{bmatrix} -\sin(\theta - \frac{2\pi}{3}) & \sin\theta \\ \\ -\sin\theta & \sin(\theta + \frac{2\pi}{3}) \end{bmatrix} \end{cases} \tag{1.33}$$

An interesting particularity of matrixes A and B has been noticed, such that:

$$AB = \frac{3}{4} I_2 \tag{1.34}$$

which leads to:

$$\begin{cases} A^{-1} = \frac{4}{3} B \\ \\ B^{-1} = \frac{4}{3} A \end{cases} \tag{1.35}$$

The inversion of matrix L_1 could be proceeded as follows:

$$\begin{cases} \Phi_{1s} = l_s\, I_{1s} & + \sqrt{3} M_{rs} A\, I_{1r} \\ \\ \Phi_{1r} = \sqrt{3} M_{rs} B\, I_{1s} + l_r\, I_{1r} \end{cases} \tag{1.36}$$

which gives:

$$\begin{cases} \dfrac{1}{\sqrt{3}M_{rs}} A^{-1}\, \Phi_{1s} = \dfrac{l_s}{\sqrt{3}M_{rs}} A^{-1}\, I_{1s} + I_{1r} \\ \\ \dfrac{1}{l_r}\, \Phi_{1r} = \dfrac{\sqrt{3}M_{rs}}{l_r} B\, I_{1s} + I_{1r} \end{cases} \tag{1.37}$$

By eliminating I_{1r} and taking into account expressions (1.35), the stator current vector I_{1s} can expressed as:

$$I_{1s} = \frac{1}{\sigma l_s} \Phi_{1s} - \frac{2}{\sqrt{3}} \frac{M}{\sigma l_s l_r} A\, \Phi_{1r} \tag{1.38}$$

where:

$$\begin{cases} M = \frac{3}{2} M_{rs} \\ \sigma = 1 - \frac{M^2}{l_s l_r} \end{cases} \quad (1.39)$$

Similarly, the rotor current vector I_{1r} can expressed as:

$$I_{1r} = -\frac{2}{\sqrt{3}} \frac{M}{\sigma l_s l_r} B \, \Phi_{1s} + \frac{1}{\sigma l_r} \Phi_{1r} \quad (1.40)$$

Accounting for Eqs. (1.38) and (1.40), one can establish the following:

$$\begin{bmatrix} I_{1s} \\ I_{1r} \end{bmatrix} = \begin{bmatrix} \frac{1}{\sigma l_s} \mathcal{I}_2 & -\frac{2}{\sqrt{3}} \frac{M}{\sigma l_s l_r} A \\ -\frac{2}{\sqrt{3}} \frac{M}{\sigma l_s l_r} B & \frac{1}{\sigma l_r} \mathcal{I}_2 \end{bmatrix} \begin{bmatrix} \Phi_{1s} \\ \Phi_{1r} \end{bmatrix} \quad (1.41)$$

The substitution of matrixes A and B by their expressions given in (1.33) leads to L_1^{-1}, such that [2]:

$$L_1^{-1} = \frac{2}{\sqrt{3}} \frac{M}{\sigma l_s l_r} \begin{bmatrix} \frac{\sqrt{3}}{2} \frac{l_r}{M} & 0 & -\sin(\theta + \frac{2\pi}{3}) & \sin\theta \\ 0 & \frac{\sqrt{3}}{2} \frac{l_r}{M} & -\sin\theta & \sin(\theta - \frac{2\pi}{3}) \\ \sin(\theta - \frac{2\pi}{3}) & -\sin\theta & \frac{\sqrt{3}}{2} \frac{l_s}{M} & 0 \\ \sin\theta & -\sin(\theta + \frac{2\pi}{3}) & 0 & \frac{\sqrt{3}}{2} \frac{l_s}{M} \end{bmatrix} \quad (1.42)$$

1.5 *Park* Transform

It consists in substituting the three stator phases and the three rotor ones by two pairs of equivalent orthogonal circuits. These fictive circuits have therefore orthogonal magnetic axis: the so-called: direct (noted "d") and quadrature (noted "q") axis. Moreover, in order to account for possible unbalanced operation, a third axis orthogonal to the (d, q)-plane noted "o" could be included. The relative positions of the magnetic axis of the stator and rotor phases with respect to those of the dqo-frame are illustrated in Fig. 1.2.

Fig. 1.2 Relative positions
of the magnetic axis of the
stator and rotor phases with
respect to those of the
dqo-frame

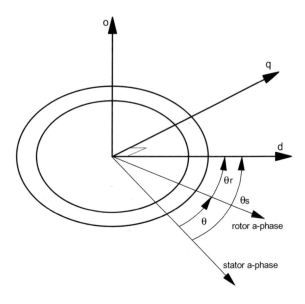

The *Park* transform enables the expression of the a-b-c components of a vector X in terms of its equivalent components expressed in the dqo-frame, such that [1, 3, 4]:

$$X_{abc} = P(\beta)X_{dqo} \tag{1.43}$$

where:

$$P(\beta) = \sqrt{\frac{2}{3}} \begin{bmatrix} \cos\beta & -\sin\beta & \frac{1}{\sqrt{2}} \\ \cos(\beta - \frac{2\pi}{3}) & -\sin(\beta - \frac{2\pi}{3}) & \frac{1}{\sqrt{2}} \\ \cos(\beta + \frac{2\pi}{3}) & -\sin(\beta + \frac{2\pi}{3}) & \frac{1}{\sqrt{2}} \end{bmatrix} \tag{1.44}$$

with:

- $\beta = \theta_s$ in the case of a vector of stator variables,
- $\beta = \theta_r$ in the case of a vector of rotor variables.

Inversely, one can express the following relation:

$$X_{dqo} = P^{-1}(\beta)X_{abc} \tag{1.45}$$

where:

$$P^{-1}(\beta) = \sqrt{\frac{2}{3}} \begin{bmatrix} \cos\beta & \cos(\beta - \frac{2\pi}{3}) & \cos(\beta + \frac{2\pi}{3}) \\ -\sin\beta & -\sin(\beta - \frac{2\pi}{3}) & -\sin(\beta + \frac{2\pi}{3}) \\ \frac{1}{\sqrt{2}} & \frac{1}{\sqrt{2}} & \frac{1}{\sqrt{2}} \end{bmatrix} \quad (1.46)$$

The choice of the dqo-frame could be arbitrary. However, and for the sake of simplification of the machine *Park* model, the following cases are commonly adopted:

- the dqo-frame is linked to the stator,
- the dqo-frame is linked to the rotor,
- the dqo-frame is linked to the rotating field.

1.5.1 Case of a dqo-Frame Linked to the Rotating Field

The dqo-frame is rotating at the synchronous speed Ω_s, which leads to:

$$\theta_s = \omega_s t \quad (1.47)$$

Accounting for the synchronism condition illustrated in Fig. 1.2, one can determine angle θ_r required for the *Park* transform of the rotor variables, as:

$$\theta_r = \theta_s - \theta \quad (1.48)$$

Knowing:

$$\theta = p\theta_m = p\Omega_m t = \omega_m t \quad (1.49)$$

Accounting for expressions (1.47) and (1.49), Eq. (1.50) gives:

$$\theta_r = (\omega_s - p\Omega_m)t = (\omega_s - \omega_m)t = \omega_r t \quad (1.50)$$

The case of a dqo-frame linked to the rotating field is systematically considered in the synthesis of the IM field-oriented control (also named vector control) strategies. Of particular interest is the dqo-frame linked to the rotating field with its d-axis aligned on the rotor flux vector. Within such a dqo-frame, the IM model turns to be similar to the one of a separately excited DC machine.

Furthermore, the dqo-frame linked to the rotating field with the d-axis aligned with the field one is exclusively used in the *Park* model of the synchronous machine.

1.5.2 Case of a dqo-Frame Linked to the Stator

In this case, the electric angles fulfil the following relations:

$$\begin{cases} \theta_s = 0 \\ \theta_r = -\theta \end{cases} \tag{1.51}$$

Thus, the inverse matrix used to transform the a-b-c stator variables turns to be:

$$P_s^{-1}(\beta = 0) = \sqrt{\frac{2}{3}} \begin{bmatrix} 1 & -\frac{1}{2} & -\frac{1}{2} \\ 0 & \frac{\sqrt{3}}{2} & -\frac{\sqrt{3}}{2} \\ \frac{1}{\sqrt{2}} & \frac{1}{\sqrt{2}} & \frac{1}{\sqrt{2}} \end{bmatrix} \tag{1.52}$$

Such a matrix is the inverse of the one introduced by *Concordia*. Furthermore, it represents the inverse of the matrix proposed by *Clarke* within the factor $\sqrt{\frac{2}{3}}$ [1]. In this case, the dqo-frame is commonly named $\alpha\beta$o-frame.

The inverse matrix used to transform the a-b-c rotor variables is expressed in terms of the electrical position, as follows:

$$P^{-1}(\beta) = \sqrt{\frac{2}{3}} \begin{bmatrix} \cos\theta & \cos(\theta + \frac{2\pi}{3}) & \cos(\theta - \frac{2\pi}{3}) \\ \sin\theta & \sin(\theta + \frac{2\pi}{3}) & \sin(\theta - \frac{2\pi}{3}) \\ \frac{1}{\sqrt{2}} & \frac{1}{\sqrt{2}} & \frac{1}{\sqrt{2}} \end{bmatrix} \tag{1.53}$$

The case of a dqo-frame linked to the stator is commonly used in the synthesis of the direct torque control (DTC) strategies dedicated to induction machines. In these strategies, the adopted formulations are generally limited to the stator voltage equations and the electromagnetic torque.

1.5.3 Case of a dqo-Frame Linked to the Rotor

In this case, the electric angles fulfil the following relations:

$$\begin{cases} \theta_s = \theta \\ \theta_r = 0 \end{cases} \tag{1.54}$$

Thus, the inverse matrix used to transform the a-b-c stator variables turns to be:

$$P_s^{-1}(\beta = \theta) = \sqrt{\frac{2}{3}} \begin{bmatrix} \cos\theta & \cos(\theta - \frac{2\pi}{3}) & \cos(\theta + \frac{2\pi}{3}) \\ -\sin\theta & -\sin(\theta - \frac{2\pi}{3}) & -\sin(\theta + \frac{2\pi}{3}) \\ \frac{1}{\sqrt{2}} & \frac{1}{\sqrt{2}} & \frac{1}{\sqrt{2}} \end{bmatrix} \tag{1.55}$$

Such a matrix is systematically considered for the transformation of the armature electrical variables of the synchronous machine. This later is characterized by a synchronous rotation of the rotor and the rotating field. In such a case, the d-axis of the dqo-frame is hold by the magnetic axis of the synchronous machine excitation source (field or permanent magnets).

The inverse matrix used to transform the a-b-c rotor variables of an induction machine is expressed as follows:

$$P_s^{-1}(\beta = 0) = \sqrt{\frac{2}{3}} \begin{bmatrix} 1 & -\frac{1}{2} & -\frac{1}{2} \\ 0 & \frac{\sqrt{3}}{2} & -\frac{\sqrt{3}}{2} \\ \frac{1}{\sqrt{2}} & \frac{1}{\sqrt{2}} & \frac{1}{\sqrt{2}} \end{bmatrix} \tag{1.56}$$

1.6 IM *Park* Model

The application of the *Park* transform to the IM a-b-c model yields its *Park* one.

1.6.1 Magnetic Equation

The substitution of the a-b-c variables by their equivalent dqo ones in Eq. (1.11) gives:

$$
\begin{bmatrix} P(\theta_s)\phi_{dqos} \\ P(\theta_r)\phi_{dqor} \end{bmatrix} = \begin{bmatrix} L_{ss} & L_{sr} \\ L_{rs} & L_{rr} \end{bmatrix} \begin{bmatrix} P(\theta_s)I_{dqos} \\ P(\theta_r)I_{dqor} \end{bmatrix}
\tag{1.57}
$$

which can be rewritten as:

$$
\begin{bmatrix} \phi_{dqos} \\ \phi_{dqor} \end{bmatrix} = \begin{bmatrix} P^{-1}(\theta_s) & \mathcal{O}_3 \\ \mathcal{O}_3 & P^{-1}(\theta_r) \end{bmatrix} \begin{bmatrix} L_{ss}P(\theta_s) & L_{sr}P(\theta_r) \\ L_{rs}P(\theta_s) & L_{rr}P(\theta_r) \end{bmatrix} \begin{bmatrix} I_{dqos} \\ I_{dqor} \end{bmatrix}
\tag{1.58}
$$

which can be rewritten as:

$$
\begin{cases} \phi_{dqos} = P^{-1}(\theta_s)L_{ss}P(\theta_s)I_{dqos} + P^{-1}(\theta_s)L_{sr}P(\theta_r)I_{dqor} \\ \phi_{dqor} = P^{-1}(\theta_r)L_{rs}P(\theta_s)I_{dqos} + P^{-1}(\theta_r)L_{rr}P(\theta_r)I_{dqor} \end{cases}
\tag{1.59}
$$

Omitting the homopolar components and regardless the dqo-frame, the development of the different matrixes gives:

$$
\begin{cases} P^{-1}(\theta_s)L_{ss}P(\theta_s) = \begin{bmatrix} l_s & 0 \\ 0 & l_s \end{bmatrix} \\[2mm] P^{-1}(\theta_s)L_{sr}P(\theta_r) = \begin{bmatrix} M & 0 \\ 0 & M \end{bmatrix} \\[2mm] P^{-1}(\theta_r)L_{rs}P(\theta_s) = \begin{bmatrix} M & 0 \\ 0 & M \end{bmatrix} \\[2mm] P^{-1}(\theta_r)L_{rr}P(\theta_r) = \begin{bmatrix} l_r & 0 \\ 0 & l_r \end{bmatrix} \end{cases}
\tag{1.60}
$$

The development of Eq. (1.59) taking into account system (1.60) yields:

• Stator:

$$
\begin{cases} \phi_{ds} = l_s i_{ds} + M i_{dr} \\ \phi_{qs} = l_s i_{qs} + M i_{qr} \end{cases}
\tag{1.61}
$$

• Rotor:

$$\begin{cases} \phi_{dr} = l_r i_{dr} + M i_{ds} \\ \phi_{qr} = l_r i_{qr} + M i_{qs} \end{cases} \qquad (1.62)$$

Once again, it should be underlined that the above flux expressions are independent of the dqo-frame.

1.6.2 Electrical Equation

The substitution of the a-b-c variables by their equivalent dqo ones in Eq. (1.15) yields:

$$\begin{bmatrix} P(\theta_s)V_{dqos} \\ P(\theta_r)V_{dqor} \end{bmatrix} = \begin{bmatrix} r_s \mathcal{I}_3 & \mathcal{O}_3 \\ \mathcal{O}_3 & r_r \mathcal{I}_3 \end{bmatrix} \begin{bmatrix} P(\theta_s)I_{dqos} \\ P(\theta_r)I_{dqor} \end{bmatrix} + \frac{d}{dt}\left\{ \begin{bmatrix} L_{ss} & L_{sr} \\ L_{rs} & L_{rr} \end{bmatrix} \begin{bmatrix} P(\theta_s)I_{dqos} \\ P(\theta_r)I_{dqor} \end{bmatrix} \right\} \qquad (1.63)$$

which can be rewritten as:

$$\begin{bmatrix} V_{dqos} \\ V_{dqor} \end{bmatrix} = \begin{bmatrix} P^{-1}(\theta_s)r_s\mathcal{I}_3 P(\theta_s) & \mathcal{O}_3 \\ \mathcal{O}_3 & P^{-1}(\theta_r)r_r\mathcal{I}_3 P(\theta_r) \end{bmatrix} \begin{bmatrix} I_{dqos} \\ I_{dqor} \end{bmatrix}$$

$$(1.64)$$

$$+ \begin{bmatrix} P^{-1}(\theta_s) & \mathcal{O}_3 \\ \mathcal{O}_3 & P^{-1}(\theta_r) \end{bmatrix} \frac{d}{dt}\left\{ \begin{bmatrix} L_{ss}P(\theta_s) & L_{sr}P(\theta_r) \\ L_{rs}P(\theta_s) & L_{rr}P(\theta_r) \end{bmatrix} \begin{bmatrix} I_{dqos} \\ I_{dqor} \end{bmatrix} \right\}$$

which can be rewritten as:

$$\begin{cases} V_{dqos} = P^{-1}(\theta_s)r_s\mathcal{I}_3 P(\theta_s)I_{dqos} + P^{-1}(\theta_s)\frac{d}{dt}\left\{ L_{ss}P(\theta_s)I_{dqos} + L_{sr}P(\theta_r)I_{dqor} \right\} \\ \\ V_{dqor} = P^{-1}(\theta_r)r_r\mathcal{I}_3 P(\theta_r)I_{dqor} + P^{-1}(\theta_r)\frac{d}{dt}\left\{ L_{rs}P(\theta_s)I_{dqos} + L_{rr}P(\theta_r)I_{dqor} \right\} \end{cases} \qquad (1.65)$$

where:

$$\begin{cases} P^{-1}(\theta_s)r_s\mathcal{I}_3 P(\theta_s) = r_s\mathcal{I}_3 \\ P^{-1}(\theta_r)r_r\mathcal{I}_3 P(\theta_r) = r_r\mathcal{I}_3 \end{cases} \qquad (1.66)$$

However, the matrix expressions $P^{-1}(\theta_s)\frac{d}{dt}\left\{L_{ss}P(\theta_s)I_{dqos}+L_{sr}P(\theta_r)I_{dqor}\right\}$ and $P^{-1}(\theta_r)\frac{d}{dt}\left\{L_{rs}P(\theta_s)I_{dqos}+L_{rr}P(\theta_r)I_{dqor}\right\}$ depend on the selected dqo-frame. Following their development, the electrical Eq. (1.65) gives:

- Stator:

$$
\begin{cases}
v_{ds} = \left(r_s + l_s\frac{d}{dt}\right)i_{ds} - l_si_{qs}\frac{d\theta_s}{dt} & + M\frac{di_{dr}}{dt} & - Mi_{qr}\frac{d\theta_s}{dt} \\[2mm]
v_{qs} = l_si_{ds}\frac{d\theta_s}{dt} & + \left(r_s + l_s\frac{d}{dt}\right)i_{qs} + Mi_{dr}\frac{d\theta_s}{dt} + M\frac{di_{qr}}{dt}
\end{cases}
\tag{1.67}
$$

- Rotor:

$$
\begin{cases}
v_{dr} = M\frac{di_{ds}}{dt} & - Mi_{qs}\frac{d\theta_r}{dt} + \left(r_r + l_r\frac{d}{dt}\right)i_{dr} - l_ri_{qr}\frac{d\theta_r}{dt} \\[2mm]
v_{qr} = Mi_{ds}\frac{d\theta_r}{dt} + M\frac{di_{qs}}{dt} & + l_ri_{dr}\frac{d\theta_r}{dt} & + \left(r_r + l_r\frac{d}{dt}\right)i_{qr}
\end{cases}
\tag{1.68}
$$

where v_{dr} and v_{qr} are null in the case of short-circuited phases in the rotor.

1.6.2.1   Case of a dqo-Frame Linked to the Rotating Field

In this case, the following relations are fulfilled:

$$
\begin{cases}
\dfrac{d\theta_s}{dt} = \omega_s \\[3mm]
\dfrac{d\theta_r}{dt} = \omega_r
\end{cases}
\tag{1.69}
$$

The electrical Eqs. (1.67) and (1.68) turn to be:

- Stator:

$$
\begin{cases}
v_{ds} = \left(r_s + l_s\frac{d}{dt}\right)i_{ds} - l_s\omega_si_{qs} & + M\frac{di_{dr}}{dt} - M\omega_si_{qr} \\[2mm]
v_{qs} = l_s\omega_si_{ds} & + \left(r_s + l_s\frac{d}{dt}\right)i_{qs} + M\omega_si_{dr} + M\frac{di_{qr}}{dt}
\end{cases}
\tag{1.70}
$$

- Rotor:

$$
\begin{cases}
v_{dr} = M\frac{di_{ds}}{dt} - M\omega_ri_{qs} + \left(r_r + l_r\frac{d}{dt}\right)i_{dr} - l_r\omega_ri_{qr} \\[2mm]
v_{qr} = M\omega_ri_{ds} + M\frac{di_{qs}}{dt} + l_r\omega_ri_{dr} & + \left(r_r + l_r\frac{d}{dt}\right)i_{qr}
\end{cases}
\tag{1.71}
$$

In the electrical Eqs. (1.160) and (1.150), the phase terminal voltages are totally expressed in terms of the phase currents. They can be expressed in a mixture current-fluxes taking into account relations (1.61) and (1.62), resulting in:

- Stator:

$$\begin{cases} v_{ds} = r_s i_{ds} + \dfrac{d\phi_{ds}}{dt} - \omega_s \phi_{qs} \\ v_{qs} = r_s i_{qs} + \dfrac{d\phi_{qs}}{dt} + \omega_s \phi_{ds} \end{cases} \tag{1.72}$$

- Rotor:

$$\begin{cases} v_{dr} = r_r i_{dr} + \dfrac{d\phi_{dr}}{dt} - \omega_r \phi_{qr} \\ v_{qr} = r_r i_{qr} + \dfrac{d\phi_{qr}}{dt} + \omega_r \phi_{dr} \end{cases} \tag{1.73}$$

Furthermore, relations (1.61) and (1.62) allow the expression of the currents in terms of the fluxes as:

- Stator:

$$\begin{cases} i_{ds} = \dfrac{l_r \phi_{ds} - M \phi_{dr}}{l_s l_r - M^2} = \dfrac{1}{l_s \sigma} \phi_{ds} - \dfrac{M}{l_s l_r \sigma} \phi_{dr} \\ i_{qs} = \dfrac{l_r \phi_{qs} - M \phi_{qr}}{l_s l_r - M^2} = \dfrac{1}{l_s \sigma} \phi_{qs} - \dfrac{M}{l_s l_r \sigma} \phi_{qr} \end{cases} \tag{1.74}$$

- Rotor

$$\begin{cases} i_{dr} = \dfrac{l_s \phi_{dr} - M \phi_{ds}}{l_s l_r - M^2} = \dfrac{1}{l_r \sigma} \phi_{dr} - \dfrac{M}{l_s l_r \sigma} \phi_{ds} \\ i_{qr} = \dfrac{l_s \phi_{qr} - M \phi_{qs}}{l_s l_r - M^2} = \dfrac{1}{l_r \sigma} \phi_{qr} - \dfrac{M}{l_s l_r \sigma} \phi_{qs} \end{cases} \tag{1.75}$$

The substitution of the currents in Eqs. (1.72) and (1.73), by their expressions in terms of the fluxes given by relations (1.74) and (1.75), enables a fully flux formulation of the phase terminal voltages, as follows:

- Stator:

$$\begin{cases} v_{ds} = \left(\dfrac{r_s}{l_s \sigma} + \dfrac{d}{dt} \right) \phi_{ds} - \omega_s \phi_{qs} \qquad\qquad - \dfrac{r_s M}{l_s l_r \sigma} \phi_{dr} \\ v_{qs} = \omega_s \phi_{ds} \qquad\qquad + \left(\dfrac{r_s}{l_s \sigma} + \dfrac{d}{dt} \right) \phi_{qs} - \dfrac{r_s M}{l_s l_r \sigma} \phi_{qr} \end{cases} \tag{1.76}$$

- Rotor:

$$\begin{cases} v_{dr} = -\dfrac{r_r M}{l_s l_r \sigma} \phi_{ds} + \left(\dfrac{r_r}{l_r \sigma} + \dfrac{d}{dt} \right) \phi_{dr} - \omega_r \phi_{qr} \\[4mm] v_{qr} = -\dfrac{r_r M}{l_s l_r \sigma} \phi_{qs} + \left(\dfrac{r_r}{l_r \sigma} + \dfrac{d}{dt} \right) \phi_{qr} + \omega_r \phi_{dr} \end{cases} \qquad (1.77)$$

1.6.2.2   Case of a dqo-Frame Linked to the Stator

In this case, the following relations are fulfilled:

$$\begin{cases} \dfrac{d\theta_s}{dt} = 0 \\[4mm] \dfrac{d\theta_r}{dt} = -\omega_m \end{cases} \qquad (1.78)$$

The electrical Eqs. (1.67) and (1.68) turn to be:

- Stator:

$$\begin{cases} v_{ds} = \left(r_s + l_s \dfrac{d}{dt} \right) i_{ds} + M \dfrac{d i_{dr}}{dt} \\[4mm] v_{qs} = \left(r_s + l_s \dfrac{d}{dt} \right) i_{qs} + M \dfrac{d i_{qr}}{dt} \end{cases} \qquad (1.79)$$

- Rotor:

$$\begin{cases} v_{dr} = M \dfrac{d i_{ds}}{dt} \quad + M \omega_m i_{qs} + \left(r_r + l_r \dfrac{d}{dt} \right) i_{dr} + l_r \omega_m i_{qr} \\[4mm] v_{qr} = -M \omega_m i_{ds} + M \dfrac{d i_{qs}}{dt} \quad - l_r \omega_m i_{dr} \qquad + \left(r_r + l_r \dfrac{d}{dt} \right) i_{qr} \end{cases} \qquad (1.80)$$

Accounting for relations (1.61) and (1.62), Eqs. (1.79) and (1.80) are rewritten as follows:

- Stator:

$$\begin{cases} v_{ds} = r_s i_{ds} + \dfrac{d\phi_{ds}}{dt} \\[4mm] v_{qs} = r_s i_{qs} + \dfrac{d\phi_{qs}}{dt} \end{cases} \qquad (1.81)$$

- Rotor:

$$\begin{cases} v_{dr} = r_r i_{dr} + \dfrac{d\phi_{dr}}{dt} + \omega_m \phi_{qr} \\[4mm] v_{qr} = r_r i_{qr} - \omega_m \phi_{dr} + \dfrac{d\phi_{qr}}{dt} \end{cases} \qquad (1.82)$$

1.6.2.3 Case of a dqo-Frame Linked to the Rotor

In this case, the following relations are fulfilled:

$$\begin{cases} \dfrac{d\theta_s}{dt} = \omega_m \\[4mm] \dfrac{d\theta_r}{dt} = 0 \end{cases} \qquad (1.83)$$

The electrical Eqs. (1.67) and (1.68) turn to be:

- Stator:

$$\begin{cases} v_{ds} = \left(r_s + l_s \dfrac{d}{dt} \right) i_{ds} - l_s \omega_m i_{qs} + M \dfrac{d i_{dr}}{dt} - M \omega_m i_{qr} \\[4mm] v_{qs} = l_s \omega_m i_{ds} + \left(r_s + l_s \dfrac{d}{dt} \right) i_{qs} + M \omega_m i_{dr} + M \dfrac{d i_{qr}}{dt} \end{cases} \qquad (1.84)$$

- Rotor:

$$\begin{cases} v_{dr} = M \dfrac{d i_{ds}}{dt} + \left(r_r + l_r \dfrac{d}{dt} \right) i_{dr} \\[4mm] v_{qr} = M \dfrac{d i_{qs}}{dt} + \left(r_r + l_r \dfrac{d}{dt} \right) i_{qr} \end{cases} \qquad (1.85)$$

Accounting for relations (1.61) and (1.62), Eqs. (1.84) and (1.85) are rewritten as:

- Stator:

$$\begin{cases} v_{ds} = r_s i_{ds} + \dfrac{d\phi_{ds}}{dt} - \omega_m \phi_{qs} \\[4mm] v_{qs} = r_s i_{qs} + \omega_m \phi_{ds} + \dfrac{d\phi_{qs}}{dt} \end{cases} \qquad (1.86)$$

- Rotor:

$$\begin{cases} v_{dr} = r_r i_{dr} + \dfrac{d\phi_{dr}}{dt} \\[4mm] v_{qr} = r_r i_{qr} + \dfrac{d\phi_{qr}}{dt} \end{cases} \qquad (1.87)$$

1.6.3 Mechanical Equation

The electromagnetic torque is the only feature which is affected by the *Park* transform. The expression of T_{em} is obtained by substituting the stator and rotor a-b-c currents by their equivalent in the dqo-frame, in Eq. (1.21), as follows:

$$T_{em} = p \left(P(\theta_r) I_{dqor} \right)^t \left\{ \frac{d}{d\theta} [L_{rs}] \right\} P(\theta_s) I_{dqos} \tag{1.88}$$

which can be rewritten as:

$$T_{em} = p I_{dqor}^t P^t(\theta_r) \left\{ \frac{d}{d\theta} [L_{rs}] \right\} P(\theta_s) I_{dqos} \tag{1.89}$$

Knowing that:

$$P^t(\theta_r) \left\{ \frac{d}{d\theta} [L_{rs}] \right\} P(\theta_s) = M \begin{bmatrix} 0 & 1 & 0 \\ -1 & 0 & 0 \\ 0 & 0 & 0 \end{bmatrix} \tag{1.90}$$

the expression of the electromagnetic torque is then:

$$T_{em} = pM \left(i_{dr} i_{qs} - i_{qr} i_{ds} \right) \tag{1.91}$$

In the case of mixed current-flux model, the electromagnetic torque expression (1.92) can be rewritten, for instance, in terms of the d-q components of the stator current and the rotor flux taking into consideration Eqs. (1.61) and (1.62), as follows:

$$T_{em} = pM \left(\left[\frac{\phi_{dr} - M i_{ds}}{l_r} \right] i_{qs} - \left[\frac{\phi_{qr} - M i_{qs}}{l_r} \right] i_{ds} \right) \tag{1.92}$$

which gives:

$$T_{em} = p \frac{M}{l_r} \left(\phi_{dr} i_{qs} - \phi_{qr} i_{ds} \right) \tag{1.93}$$

This expression is considered in the field-oriented control of the induction machine with the d-axis aligned with the rotor flux vector. Hence, the quadrature components ϕ_{qr} is null, and the electromagnetic torque expression is reduced to:

$$T_{em} = p \frac{M}{l_r} \phi_{dr} i_{qs} \tag{1.94}$$

which is similar to the one of a DC machine.

Similar developments have led to the following flux-current expressions of the electromagnetic torque:

$$
\begin{cases}
T_{em} = p\dfrac{M}{T_s}\ (\phi_{qs}i_{dr} - \phi_{ds}i_{qr}) \\[2mm]
T_{em} = p\quad (\phi_{ds}i_{qs} - \phi_{qs}i_{ds}) \\[2mm]
T_{em} = p\quad (\phi_{qr}i_{dr} - \phi_{dr}i_{qr})
\end{cases}
\tag{1.95}
$$

1.7 Park Model-Based Analysis of the IM Steady-State Operation

Let us assume that the IM has reached a steady-state operation characterized by sinusoidal stator and rotor variables, and therefore a given slip.

Let us define a phasor $\overline{X}$, such that:

$$
\overline{X} = x_d + jx_q
\tag{1.96}
$$

where x_d and x_q are the direct and quadrature components of a vector X of a-b-c electric variables (voltage, current, or flux) corresponding to $\overline{X}$, such that:

$$
\begin{cases}
x_a = \sqrt{2}X_{rms} \cos \omega t - \phi) \\[2mm]
x_b = \sqrt{2}X_{rms} \cos(\omega t - \phi - \dfrac{2\pi}{3}) \\[2mm]
x_c = \sqrt{2}X_{rms} \cos(\omega t - \phi + \dfrac{2\pi}{3})
\end{cases}
\tag{1.97}
$$

where X_{rms} is the root mean square of X and where $\omega = \omega_s$ for the stator variables and $\omega = \omega_r$ for the rotor ones.

The application of the *Park* transform whose dq-frame (instead of dqo-frame because the hompolar component is null in the case of balanced a-b-c components) is rotating at the angular frequency ω with a null initial phase leads to:

$$
\begin{cases}
x_d = \sqrt{3}X_{rms} \cos \phi \\[2mm]
x_q = -\sqrt{3}X_{rms} \sin \phi
\end{cases}
\tag{1.98}
$$

The magnitude (or modulus) of $\overline{X}$ is expressed as:

$$
\|\overline{X}\| = \sqrt{x_d^2 + x_q^2} = \sqrt{3}X_{rms}
\tag{1.99}
$$

1.7.1 Steady-State Stator Current Formulation

Under steady-state operation, the rotor voltage equations could be regrouped in a phasor form, regardless the considered dqo-frame. In the case of a full current dqo model of the IM, the resulting rotor voltage equation is as follows:

$$\overline{V}_r = 0 = (r_r + jl_r\omega_r)\overline{I}_r + jM\omega_r\overline{I}_s \tag{1.100}$$

which leads to the expression of the rotor current phasor $\overline{I}_r$, such that:

$$\overline{I}_r = -\frac{jM\omega_r}{(r_r + jl_r\omega_r)}\overline{I}_s \tag{1.101}$$

Similarly, the stator flux phasor could be expressed as:

$$\overline{\Phi}_s = l_s\overline{I}_s + M\overline{I}_r \tag{1.102}$$

that gives:

$$\overline{I}_r = \frac{1}{M}\left(\overline{\Phi}_s - l_s\overline{I}_s\right) \tag{1.103}$$

The equality of both rotor current phasors $\overline{I}_r$, given by Eqs. (1.103) and (1.100), yields the stator current phasor $\overline{I}_s$, as follows:

$$\overline{I}_s = \frac{1}{l_s}\left(\frac{r_r + jl_r\omega_r}{r_r + j\sigma l_r\omega_r}\right)\overline{\Phi}_s \tag{1.104}$$

and then:

$$I_{s-rms} = \frac{1}{l_s}\sqrt{\frac{1+\left(\frac{l_r\omega_r}{r_r}\right)^2}{1+\left(\frac{\sigma l_r\omega_r}{r_r}\right)^2}}\,\Phi_{s-rms} \tag{1.105}$$

Equation (1.105) represents the basis on which are founded the scalar control strategies of the IM fed by current source inverters. Figure 1.3 shows the variations of I_{s-rms} with respect to the rotor frequency f_r, assuming a constant Φ_{s-rms}.

1.7.2 Steady-State Stator Voltage Formulation

Under steady-state operation, the stator voltage equations could be regrouped in a phasor form, regardless the considered dqo-frame, as:

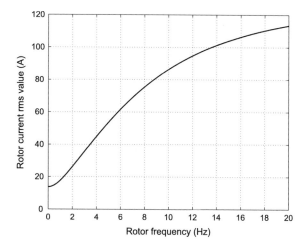

Fig. 1.3 I_{s-rms} versus f_r, assuming a constant Φ_{s-rms}

$$\overline{V}_s = (r_s + jl_s\omega_s)\overline{I}_s + jM\omega_s\overline{I}_r \qquad (1.106)$$

Accounting for Eqs. (1.101) et (1.106), the stator voltage phasor $\overline{V}_s$ could be expressed in terms of $\overline{I}_s$ as follows:

$$\overline{V}_s = \left(r_s + jl_s\omega_s + \frac{M^2\omega_r\omega_s}{(r_r + jl_r\omega_r)} \right) \overline{I}_s \qquad (1.107)$$

The substitution of the stator current phasor $\overline{I}_s$, by its expression (1.104) in Eq. (1.107), gives:

$$\overline{V}_s = \left(r_s + jl_s\omega_s + \frac{M^2\omega_r\omega_s}{(r_r + jl_r\omega_r)} \right) \frac{1}{l_s} \left(\frac{r_r + jl_r\omega_r}{r_r + j\sigma l_r\omega_r} \right) \overline{\Phi}_s \qquad (1.108)$$

The development of Eq. (1.108) enables a formulation of the stator voltage phasor $\overline{V}_s$ in terms of the stator flux phasor $\overline{\Phi}_s$, the stator angular frequency ω_s, and the rotor one ω_r, as follows:

$$\overline{V}_s = \frac{1}{l_s} \left(\frac{(r_s + jl_s\omega_s)(r_r + jl_r\omega_r) + M^2\omega_s\omega_r}{r_r + j\sigma l_r\omega_r} \right) \overline{\Phi}_s \qquad (1.109)$$

then:

$$\overline{V}_s = \frac{r_s}{l_s} \frac{\left(1 - \sigma\frac{l_s l_r}{r_s r_r}\omega_s\omega_r \right) + j\left(\frac{l_s}{r_s}\omega_s + \frac{l_r}{r_r}\omega_r \right)}{1 + j\frac{\sigma l_r\omega_r}{r_r}} \overline{\Phi}_s \qquad (1.110)$$

which leads to:

Fig. 1.4 V_{s-rms} versus f_r, for $f_r = 0.1$ Hz then $f_r = 1$ to 10 Hz, step 1 Hz (bottom to top), assuming a constant Φ_{s-rms}

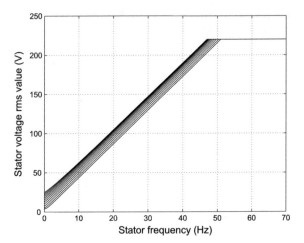

$$V_{s-rms} = \frac{r_s}{l_s} \sqrt{\frac{\left(1 - \sigma \frac{l_s l_r}{r_s r_r} \omega_s \omega_r\right)^2 + \left(\frac{l_s}{r_s} \omega_s + \frac{l_r}{r_r} \omega_r\right)^2}{1 + \left(\frac{\sigma l_r \omega_r}{r_r}\right)^2}} \; \Phi_{s-rms} \quad (1.111)$$

In the manner of Eqs. (1.105) and (1.111) represents the basis on which are founded the scalar control strategies of the IM fed by voltage source inverters. Figure 1.4 illustrates the variations of V_{s-rms} in terms of the stator frequency f_s, for different values of f_r, assuming a constant Φ_{s-rms}.

1.7.3 Steady-State Electromagnetic Torque Formulation

The electromagnetic torque expression (1.88) could be rewritten in terms of the stator and rotor current phasors, as follows [2]:

$$T_{em} = pM \, \Im m \left(\overline{I}_s \overline{I}_r^*\right) \quad (1.112)$$

Accounting for expression (1.103), one of the electromagnetic torque (1.112) is rewritten as follows:

$$T_{em} = p \, \Im m \left(\overline{I}_s \left(\overline{\Phi}_s^* - l_s \overline{I}_s^*\right)\right) = p \, \Im m \left(\overline{I}_s \overline{\Phi}_s^* - l_s \overline{I}_s \overline{I}_s^*\right) \quad (1.113)$$

Knowing that:

$$\Im m \left(\overline{I}_s \overline{I}_s^*\right) = 0 \quad (1.114)$$

expression (1.113) is reduced as follows:

$$T_{em} = p \, \Im m \left(\overline{I}_s \overline{\Phi}_s^* \right) \tag{1.115}$$

The substitution of the stator voltage phasor $\overline{I}_s$ by its expression (1.104) in the electromagnetic torque (1.115) gives:

$$T_{em} = = \frac{p}{l_s} \, \Im m \left(\frac{r_r + j l_r \omega_r}{r_r + j \sigma l_r \omega_r} \right) \| \overline{\Phi} \|^2 \tag{1.116}$$

then:

$$T_{em} = \frac{p}{l_s} \left(\frac{r_r l_r \omega_r (1 - \sigma)}{r_r^2 + (\sigma l_r \omega_r)^2} \right) \| \overline{\Phi}_s \|^2 \tag{1.117}$$

Finally, under steady-state operation and for a sinusoidal power supply connected in the stator terminals, the electromagnetic torque could be expressed as follows:

$$T_{em} = 3p \frac{M^2}{r_r l_s^2} \frac{\omega_r}{1 + \left(\frac{\sigma l_r \omega_r}{r_r} \right)^2} \Phi_{s-rms}^2 \tag{1.118}$$

Taking into account the synchronization condition:

$$\omega_r = \omega_s - p\Omega_m \tag{1.119}$$

expression (1.118) turns to be:

$$T_{em} = 3p \frac{M^2}{r_r l_s^2} \frac{(\omega_s - p\Omega_m)}{1 + \left(\frac{\sigma l_r (\omega_s - p\Omega_m)}{r_r} \right)^2} \Phi_{s-rms}^2 \tag{1.120}$$

Neglecting the voltage drop across the stator resistor, Eq. (1.106) is reduced to:

$$\overline{V}_s = j l_s \omega_s \overline{I}_s + j M \omega_s \overline{I}_r = j \omega_s \overline{\Phi}_s \tag{1.121}$$

then the electromagnetic torque expression (1.120) turns to be:

$$T_{em} = 3p \frac{M^2}{r_r (l_s \omega_s)^2} \frac{(\omega_s - p\Omega_m)}{1 + \left(\frac{\sigma l_r (\omega_s - p\Omega_m)}{r_r} \right)^2} V_{s-rms}^2 \tag{1.122}$$

Fig. 1.5 IM mechanical
characteristics (T_{em} vs. Ω_m)
for f_s = 5–50 Hz step 5 Hz,
assuming a constant Φ_{s-rms}
calculated according to
relation (1.123)

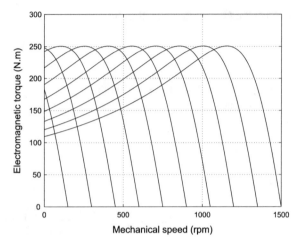

Fig. 1.6 IM mechanical
characteristics (T_{em} vs. Ω_m)
for f_s = 5–50 Hz step 5 Hz,
assuming a constant ratio
V_{s-rms}/ω_s equal to the one
of the rated point

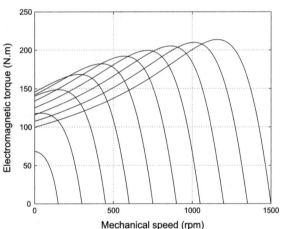

Figure 1.5 shows the mechanical characteristics (T_{em} vs. Ω_m) for different values
of f_s assuming a constant stator flux rms value, such as:

$$\Phi_{s-rms} = \left(\frac{V_{s-rms}}{\omega_s}\right)_{\text{(rated point)}} \tag{1.123}$$

Actually, the IM behaviour is different from what is shown in Fig. 1.5. Indeed,
feeding the stator terminals by voltages of variable V_{s-rms} according to ω_s in such a
way that their ratio V_{s-rms}/ω_s remains constant equal to the one of the rated point,
the mechanical characteristics turn to be as illustrated in Fig. 1.6.

1.7.4 Steady-State Powers Formulation

Basically, it is well known that the active power P is the average value of the instantaneous one $P(t)$. In polyphase circuits, and for sinusoidal voltages and currents, the following relation is fulfilled:

$$P = P(t) \tag{1.124}$$

In the case of three-phase circuits, the instantaneous active power $P(t)$ is expressed in terms of a-b-c components as follows:

$$P(t) = v_a i_a + v_b i_b + v_c i_c = V_{abc}^t I_{abc} \tag{1.125}$$

The application of the *Park* transform to Eq. (1.125) gives:

$$P(t) = \left(P(\beta)V_{dqos}\right)^t \left(P(\beta)I_{dqos}\right) = \left(V_{dqos}^t P(\beta)^t\right)\left(P(\beta)I_{dqos}\right) \tag{1.126}$$

Omitting the homopolar components, the matrix product $P(\beta)^t P(\beta)$ turns to be:

$$P(\beta)^t P(\beta) = \frac{2}{3} \begin{bmatrix} \cos\beta & \cos(\beta - \frac{2\pi}{3}) & \cos(\beta + \frac{2\pi}{3}) \\ -\sin\beta & -\sin(\beta - \frac{2\pi}{3}) & -\sin(\beta + \frac{2\pi}{3}) \end{bmatrix} \begin{bmatrix} \cos\beta & -\sin\beta \\ \cos(\beta - \frac{2\pi}{3}) & -\sin(\beta - \frac{2\pi}{3}) \\ \cos(\beta + \frac{2\pi}{3}) & -\sin(\beta + \frac{2\pi}{3}) \end{bmatrix}$$

which is equal to $\mathcal{I}_2$.

Finally, the expression of the active power P is reduced to:

$$P = P(t) = v_d i_d + v_q i_q \tag{1.127}$$

The reactive power is expressed as follows:

$$Q = \overrightarrow{V}_{abc}^t I_{abc} \tag{1.128}$$

where $\overrightarrow{V}_{abc}$ is made up of a-b-c voltages which are shifted by $-\frac{\pi}{2}$ with their respective of V_{abc}. That is to say:

$$\begin{cases} V_{abc}^t = \sqrt{2}V_{rms}\left[\cos\omega t \;\; \cos(\omega t - \frac{2\pi}{3}) \;\; \cos(\omega t + \frac{2\pi}{3})\right] \\ \overrightarrow{V}_{abc}^t = \sqrt{2}V_{rms}\left[\sin\omega t \;\; \sin(\omega t - \frac{2\pi}{3}) \;\; \sin(\omega t + \frac{2\pi}{3})\right] \end{cases} \tag{1.129}$$

A formulation similar to the one of P, where the components of V_{abc} are substituted by those of $\overrightarrow{V}_{abc}$, enabled the derivation of the reactive power Q and has led to the following:

$$Q = v_q i_d - v_d i_q \tag{1.130}$$

1.8   Case Study: Wind Energy Converters

1.8.1   Doubly Fed Operation of the IMs

From a topological point of view, the doubly fed induction machine (DFIM) is similar to the wound rotor induction machine. However, the two machines differ by their principle of operations. While the one of the wound rotor induction machine relays on the induction phenomenon presented in Sect. 1.2, the principle of operation of the DFIM is based on the synchronization of the stator and rotor rotating fields which turn to be totally independent. Indeed, two scenarios are systematically distinguished, such that:

- the stator and rotor rotating fields are turning in the same direction. In this case, their synchronization is achieved by a mechanical speed lower than the one of the stator rotating field, yielding the so-called hyposynchronism,
- the stator and rotor rotating fields are turning in opposite directions. In this case, their synchronization is achieved by a mechanical speed higher than the one of the stator rotating field, yielding the so-called hypersynchronism.

The principle of the synchronization of the stator and rotor rotating fields is illustrated in Fig. 1.7, where Ω_s, Ω_r, and Ω_m are the stator rotating field speed, the rotor rotating field speed, and the mechanical speed, respectively.

Referring to Fig. 1.7, the synchronization between the stator and rotor rotating fields is expressed as follows:

$$\Omega_m = \begin{cases} \Omega_s - \Omega_r & \text{hyposynchronism} \\ \Omega_s + \Omega_r & \text{hypersynchronism} \end{cases} \tag{1.131}$$

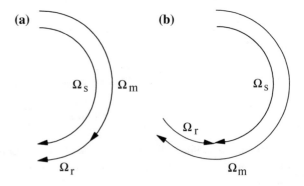

Fig. 1.7 Principle of the synchronization of the stator and rotor rotating fields in a DFIM. **Legend: a** hyposynchronism, **b** hypersynchronism

To sum up, the DFIM has a topology identical to the wound rotor induction machine and a principle of operation that represents a generalization of the synchronous machine one. Indeed, synchronous machines, to be treated in the second chapter, are considered as a particular case of the DFIM for which the speed of the rotor rotating field Ω_r is a null; a statement basically obvious in so far as the field is fed by a DC current.

1.8.2 Integration of DFIMs in Wind Generating Systems

The DFIM has been and continues to be a viable candidate in wind power generation systems. The major motivations behind the selection of DFIMs for such a sustainable application are as follows:

- the possibility of converting the wind mechanical power into electricity at variable speed which is in full harmony with the random behaviour of the wind energy,
- the stator circuits are directly connected to the grid while the rotor ones are fed by a power electronic converter through which the DFIM is controlled. It enables: (i) the synchronization of the DFIM to the grid and (ii) a flexible control of the active and reactive powers thanks to the implementation of dedicated control strategies (vector and direct power control strategies). Furthermore, it should be underlined that the power electronic converter in the rotor circuits is sized for a power flow not exceed 30% of the machine rated power. This represents a crucial-cost benefit,
- the rotor circuits are fed by the slip frequency which is limited to ±30% of the grid one. This leads to a reduction of the commutation losses within the power electronic converter in the rotor circuits, and therefore to an improved energy efficiency.

In spite of the above-listed advantages, the integration of DFIMs in wind power generating systems presents some limitations, such as:

1. the limited pole pair requires the integration of a multiplier between the shaft of the turbine and the DFIM one,
2. the brush-ring system enabling the connection to the rotor circuits needs a systematic maintenance especially in offshore installations where the corrosion affects the quality of the sliding contacts.

1.8.3 Brushless Cascaded Doubly Fed Machines

An approach to discard the second limitation of DFIMs has been proposed in [5]. It consists in the association of two wind rotor induction machines which are back-to-back connected through their rotor circuits and are mechanically coupled, yielding

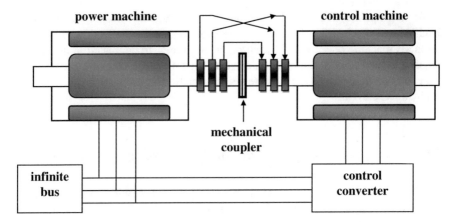

Fig. 1.8 Electrical connections of the BCDFM

the so-called brushless cascaded doubly fed machines (BCDFM). Consequently, the brush-ring system turns to be useless which improves the machine reliability.

Figure 1.8 shows a BCDFM equipping a wind generating system where the rotor circuits are interconnected with inverted phase sequences. The left-side machine is called the power machine as far as it achieves the electromechanical conversion of the wind power. While the right-side one is called the control machine. With its stator fed by a static converter, the control machine is involved in the management of the active and reactive powers between the BCDFM and the infinite bus.

1.8.3.1 Steady-State Operation

Let us call:

- Ω_m the mechanical speed of both machines,
- ω_p and ω_c the stator angular frequencies of the power and control machines, respectively,
- P_p and P_c the pole pair numbers of the power and control machines, respectively,
- s_p and s_c the slips of the power and control machines, respectively.

Question Establish the expression of Ω_m in terms of:

- ω_p, P_p, and s_p.
- ω_c, P_c, and s_c.

Answer

$$\boxed{\Omega_m = \left(1 - s_p\right) \frac{\omega_p}{P_p}} \tag{1.132}$$

and

$$\Omega_m = (1 - s_c) \frac{\omega_c}{P_c} \qquad (1.133)$$

Question Establish the expression of Ω_m in terms of ω_p, ω_c, P_p, and P_c. Conclude on the similitude of the BCDFM with one of the conventional AC machines to be identified.

Answer Accounting for the inverted phase sequences of the rotor circuits of both machines, the resulting rotating fields have opposite speeds which lead to:

$$s_p \omega_p = -s_c \omega_c \qquad (1.134)$$

From Eqs. (1.132) and (1.133), one can deduce:

$$\begin{cases} s_p \omega_p = \omega_p - P_p \Omega_m \\ s_c \omega_p = \omega_c - P_c \Omega_m \end{cases} \qquad (1.135)$$

Accounting for Eqs. (1.134) and (1.135), one can write the following equality:

$$\omega_p - P_p \Omega_m = -(\omega_c - P_c \Omega_m) \qquad (1.136)$$

that leads to:

$$\Omega_m = \frac{\omega_p + \omega_c}{P_p + P_c} \qquad (1.137)$$

Referring to expression (1.137), one can clearly deduce that the BCDFM with inverted phase sequences of the rotor circuit interconnection is equivalent to a synchronous machine that has a pole pair number equal to $(P_p + P_c)$ and an angular frequency of the armature current equal to $(\omega_p + \omega_c)$.

1.8.3.2 Application of the *Park* Transform

Let us express the rotor phase voltages of the power machine, as follows:

$$\begin{cases} v_{arp} = \sqrt{2} V_{rp} \cos(s_p \omega_p t) \\ v_{brp} = \sqrt{2} V_{rp} \cos(s_p \omega_p t - \frac{2\pi}{3}) \\ v_{crp} = \sqrt{2} V_{rp} \cos(s_p \omega_p t + \frac{2\pi}{3}) \end{cases} \qquad (1.138)$$

where V_{rp} is the rms value of the rotor phase voltages.

The resulting currents in the rotor phases of the power machine are:

$$\begin{cases} i_{arp} = \sqrt{2}I_{rp}\cos(s_p\omega_p t - \varphi_r) \\ i_{brp} = \sqrt{2}I_{rp}\cos(s_p\omega_p t - \varphi_r - \frac{2\pi}{3}) \\ i_{crp} = \sqrt{2}I_{rp}\cos(s_p\omega_p t - \varphi_r + \frac{2\pi}{3}) \end{cases} \tag{1.139}$$

where I_{rp} is the rms value of the rotor phase currents.

Let us consider the *Park* transform whose dq-frame is rotating with the rotating field created by the phase currents ($i_{arp}, i_{brp}, i_{crp}$), and whose d-axis is aligned with the phasor $\overline{V}_{rp}$ obtained following the *Park* transform of the rotor phase voltages ($v_{arp}, v_{brp}, v_{crp}$).

Furthermore, let us define the phasors $\overline{I}_{rp}$ and $\overline{I}_{rc}$ as:

$$\begin{cases} \overline{I}_{rp} = i_{drp} + ji_{qrp} \\ \overline{I}_{rc} = i_{drc} + ji_{qrc} \end{cases} \tag{1.140}$$

where i_{drp} and i_{qrp} are the direct and quadrature components obtained following the application of the *Park* transform to the rotor phase currents of the power machine, and i_{drc} and i_{qrc} are the direct and quadrature components obtained following the application of the *Park* transform to the rotor phase currents of the control machines.

Question Establish the expressions of $\overline{I}_{rp}$ and $\overline{I}_{rc}$, and the relation linking these two phasors.

Answer The *Park* inverse matrix is expressed as:

$$P^{-1}(\beta) = \sqrt{\frac{2}{3}} \begin{bmatrix} \cos\beta & \cos(\beta - \frac{2\pi}{3}) & \cos(\beta + \frac{2\pi}{3}) \\ -\sin\beta & -\sin(\beta - \frac{2\pi}{3}) & -\sin(\beta + \frac{2\pi}{3}) \end{bmatrix} \tag{1.141}$$

where $\beta = s_p\omega_p t$

Its application to the rotor phase currents of the power machine is expressed as:

$$\begin{bmatrix} i_{drp} \\ i_{qrp} \end{bmatrix} = \frac{2I_{rp}}{\sqrt{3}} \begin{bmatrix} \cos s_p\omega_p t & \cos(s_p\omega_p t - \frac{2\pi}{3}) & \cos(s_p\omega_p t + \frac{2\pi}{3}) \\ -\sin s_p\omega_p t & -\sin(s_p\omega_p t - \frac{2\pi}{3}) & -\sin(s_p\omega_p t + \frac{2\pi}{3}) \end{bmatrix} \begin{bmatrix} \cos(s_p\omega_p t - \varphi_r) \\ \cos(s_p\omega_p t - \varphi_r - \frac{2\pi}{3}) \\ \cos(s_p\omega_p t - \varphi_r + \frac{2\pi}{3}) \end{bmatrix}$$

which gives:

$$\begin{cases} i_{drp} = \sqrt{3}I_{rp}\cos\varphi_r \\ \\ i_{qrp} = -\sqrt{3}I_{rp}\sin\varphi_r \end{cases} \tag{1.142}$$

and then:

$$\overline{I}_{rp} = \sqrt{3}I_{rms}(\cos \varphi_r - j \sin \varphi_r) \tag{1.143}$$

Now let us consider, for instance, that the inverted phase sequence interconnections within the rotor circuits concern phases "b" and "c". This leads to a field rotating in the air gap of the control machine in the clockwise direction. Thus, the angle involved in the *Park* transform turns to be:

$$\beta = -s_p\omega_p t = s_c\omega_c t \tag{1.144}$$

Then, the application of the *Park* transform to the rotor phase currents of the control machine is expressed as:

$$\begin{bmatrix} i_{drc} \\ i_{qrc} \end{bmatrix} = -\frac{2I_{rp}}{\sqrt{3}} \begin{bmatrix} \cos s_p\omega_p t & \cos(s_p\omega_p t + \frac{2\pi}{3}) & \cos(s_p\omega_p t - \frac{2\pi}{3}) \\ \sin s_p\omega_p t & \sin(s_p\omega_p t + \frac{2\pi}{3}) & \sin(s_p\omega_p t - \frac{2\pi}{3}) \end{bmatrix} \begin{bmatrix} \cos(s_p\omega_p t - \varphi_r) \\ \cos(s_p\omega_p t - \varphi_r + \frac{2\pi}{3}) \\ \cos(s_p\omega_p t - \varphi_r - \frac{2\pi}{3}) \end{bmatrix}$$

which gives:

$$\begin{cases} i_{drc} = -\sqrt{3}I_{rp} \cos \varphi_r \\ i_{qrc} = -\sqrt{3}I_{rp} \sin \varphi_r \end{cases} \tag{1.145}$$

and then:

$$\overline{I}_{rc} = -(\sqrt{3}I_{rms}(\cos \varphi_r + j \sin \varphi_r)) \tag{1.146}$$

Finally, one can conclude that:

$$\overline{I}_{rc} = -\overline{I}_{rp}^* \tag{1.147}$$

Considering the same approach, one can establish the relation between the voltage phasors $\overline{V}_{rp}$ and $\overline{V}_{rc}$, as:

$$\overline{V}_{rc} = \overline{V}_{rp}^* \tag{1.148}$$

1.8.3.3 BCDFM Park Model

Let us consider the *Park* model with the dq-frame linked to the rotating field of the power machine. Accounting for the following items:

$\Rightarrow$ the induction machine full current *Park*,

$\Rightarrow$ adding subscript "p" to the power machine parameters and variables and subscript "c" to the control machine parameters and variables,

$\Rightarrow$ ω_r the angular frequency of the power machine rotor variables.

Question Establish the power machine full current *Park* model limited to the voltage equations.

Answer The electrical equations of the power machine expressed in terms of the *Park* variables are as follows:

• **Stator**:

$$
\begin{cases}
v_{dsp} = \left(r_{sp} + l_{sp}\dfrac{d}{dt}\right) i_{dsp} - l_{sp}\omega_p i_{qsp} \qquad\qquad + M_p\dfrac{di_{drp}}{dt} - M_p\omega_p i_{qrp} \\[3mm]
v_{qsp} = l_{sp}\omega_p i_{dsp} \qquad + \left(r_{sp} + l_{sp}\dfrac{d}{dt}\right) i_{qsp} + M_p\omega_p i_{drp} + M_p\dfrac{di_{qrp}}{dt}
\end{cases}
\tag{1.149}
$$

• **Rotor**:

$$
\begin{cases}
v_{drp} = M_p\dfrac{di_{dsp}}{dt} - M_p\omega_r i_{qsp} + \left(r_{rp} + l_{rp}\dfrac{d}{dt}\right) i_{drp} - l_{rp}\omega_r i_{qrp} \\[3mm]
v_{qrp} = M_p\omega_r i_{dsp} + M_p\dfrac{di_{qsp}}{dt} + l_{rp}\omega_r i_{drp} \qquad + \left(r_{rp} + l_{rp}\dfrac{d}{dt}\right) i_{qr}
\end{cases}
\tag{1.150}
$$

Question Rewrite the established model in a phasor form. Deduce the one of the control machine.

Answer Accounting for the previously established equations, one can express the power machine full current *Park* model in terms of phasor variables as follows:

• Stator of the power machine:

$$
\overline{V}_{sp} = \left(\left(r_{sp} + l_{sp}\frac{d}{dt}\right) + jl_{sp}\omega_p\right)\overline{I}_{sp} + \left(M_p\frac{d}{dt} + jM_p\omega_p\right)\overline{I}_{rp} \tag{1.151}
$$

• Rotor of the power machine:

$$
\overline{V}_{rp} = \left(M_p\frac{d}{dt} + jM_p\omega_r\right)\overline{I}_{sp} + \left(\left(r_{rp} + l_{rp}\frac{d}{dt}\right) + jl_{rp}\omega_r\right)\overline{I}_{rp} \tag{1.152}
$$

The *Park* model of the control machine expressed in terms of phasor variables can be easily deduced from the power machine one, as follows:

• Stator of the control machine:

$$
\overline{V}_{sc} = \left(\left(r_{sc} + l_{sc}\frac{d}{dt}\right) + jl_{sc}\omega_c\right)\overline{I}_{sc} + \left(M_c\frac{d}{dt} + jM_c\omega_c\right)\overline{I}_{rc} \tag{1.153}
$$

- Rotor of the control machine:

$$\overline{V}_{rc} = \left(M_c \frac{d}{dt} - jM_c\omega_r \right) \overline{I}_{sc} + \left(\left(r_{rc} + l_{rc}\frac{d}{dt} \right) - jl_{rc}\omega_r \right) \overline{I}_{rc} \quad (1.154)$$

Let us call $\overline{V}_r$ and $\overline{I}_r$ the power machine rotor phase voltage and current phasors, respectively, with:

$$\begin{cases} \overline{V}_{rp} = \overline{V}_r \\ \overline{I}_{rp} = \overline{I}_r \end{cases} \quad (1.155)$$

Relations (1.147) and (1.148) give:

$$\begin{cases} \overline{V}_{rc} = \overline{V}_r^* \\ \overline{I}_{rp} = -\overline{I}_r^* \end{cases} \quad (1.156)$$

Accounting for relations (1.155) and (1.156), the *Park* models of the power and control machines are rewritten as follows:

Power Machine Electrical Equations

- Stator:

$$\overline{V}_{sp} = \left(\left(r_{sp} + l_{sp}\frac{d}{dt} \right) + jl_{sp}\omega_p \right) \overline{I}_{sp} + \left(M_p\frac{d}{dt} + jM_p\omega_p \right) \overline{I}_r \quad (1.157)$$

- Rotor:

$$\overline{V}_r = \left(M_p\frac{d}{dt} + jM_p\omega_r \right) \overline{I}_{sp} + \left(\left(r_{rp} + l_{rp}\frac{d}{dt} \right) + jl_{rp}\omega_r \right) \overline{I}_r \quad (1.158)$$

Control Machine Electrical Equations

- Stator:

$$\overline{V}_{sc} = \left(\left(r_{sc} + l_{sc}\frac{d}{dt} \right) + jl_{sc}\omega_c \right) \overline{I}_{sc} - \left(M_c\frac{d}{dt} + jM_c\omega_c \right) \overline{I}_r^* \quad (1.159)$$

- Rotor:

$$\overline{V}_r^* = \left(M_c\frac{d}{dt} - jM_c\omega_r \right) \overline{I}_{sc} - \left(\left(r_{rc} + l_{rc}\frac{d}{dt} \right) - jl_{rc}\omega_r \right) \overline{I}_r^* \quad (1.160)$$

References

1. J. Lesenne, F. Notelet, G. Seguier, *Introduction to the Deep Electrotechnic (in French)* (Technique et Documentation, Paris, France, 1980)
2. A. Masmoudi, Contribution to the voltage control of doubly-fed machines (in French). Ph.D. Dissertation, Pierre and Marie Curie (Paris 6) University, France, Paris (1994)
3. P. Barret, M. Magnien, *Transient Regimes of Rotating Electrical Machines (in French)* (Editions Eyrolles, Paris, France, 1982)
4. J. Chatelain, *Electrical Machines* (Edition Georgi, Lausanne, Suitzerland, 1983)
5. F. Jallali, A. Masmoudi, Modeling and analysis of BCDFM: effect of rotor-phase sequence connection. Int. J. Comput. Math. Electr. Electron. Eng. **31**(1), 261–278 (2012)

Chapter 2
Synchronous Machine Modelling

Abstract The chapter deals with the modelling of the synchronous machines (SMs). Following the introduction of the continuous development of the SMs, their a-b-c model is established considering the case of salient pole machines. Then, the Park transform is applied to the established a-b-c model, leading to the Park one. A special attention is paid to the formulation and analysis of the electromagnetic torque with an investigation of the variations of its synchronizing and reluctant components in terms of the torque angle. Then, a characterization of the operation at (i) maximum torque and (ii) unity power factor is carried out before focusing the flux-weakening approaches that could be implemented in SMs considering both smooth and salient pole topologies. The chapter is achieved by a case study dealing with an investigation of the main features of the electric drive unit of a hybrid propulsion system and the possibility of their improvement with emphasis on the extension of the flux-weakening range.

Keywords Synchronous machines · Modelling · A-B-C model · *Park* model
Electromagnetic torque · Smooth/salient pole · Flux weakening

2.1 Introduction

Unlike induction machines (IMs) which have a family tree limited to two members: (i) the squirrel cage IM and (ii) the wound rotor IM, the SMs have a very ramified one. Indeed, in recent years, new concepts of SMs are more and more introduced to equip a variety of applications, covering a wide range of power. Of particular interest are permanent magnet (PM)-excited SMs, also called "brushless" SMs, which are currently given an increasing attention due to their performance and the large number of design freedom degrees.

Actually, there are different criteria based on which the SMs could be classified. The classical ones are:

© The Author(s) 2018
A. Masmoudi, *Control Oriented Modelling of AC Electric Machines*,
SpringerBriefs in Electrical and Computer Engineering,
https://doi.org/10.1007/978-981-10-9056-1_2

- The smooth pole SMs in which the air gap is assumed to be constant with the slotting effect neglected. These are characterized by the production of a synchronizing torque,
- The salient pole SMs in which the air gap presents a variable reluctance that makes the flux moving in this area within specific tubes corresponding to the "pole shoes". These SM topologies are characterized by the production of an electromagnetic torque which is the superposition of a synchronizing and a reluctant torque.

An other classical classification criterion is related to the dc excitation which could be achieved using:

- A brush-ring system,
- An exciter which enables the elimination of the brush-ring system and the associated systematic maintenance.

Finally, the most challenging classification criterion is the type of excitation:

- No excitation yielding the switched reluctance SMs which are characterized by the production of a reluctant torque,
- Field excitation,
- PM excitation which enables three types of flux paths in the air gap, yielding:

  - Radial flux machines,
  - Axial flux machines,
  - Transverse flux machines.

2.2 Synchronous Machine Modelling

2.2.1 A-B-C Model

In the manner of the IM, the a-b-c model of the SM considers the electrical variables applied to and measured in the SM circuits. These variables are illustrated in Fig. 2.1 which shows a schematic representation of a three-phase salient pole DC-excited SM.

2.2.1.1 Electrical Equation

The application of the *Ohm* law to the different circuits of the SM yields:

- Armature:

$$V = RI + \frac{d\Phi}{dt} \tag{2.1}$$

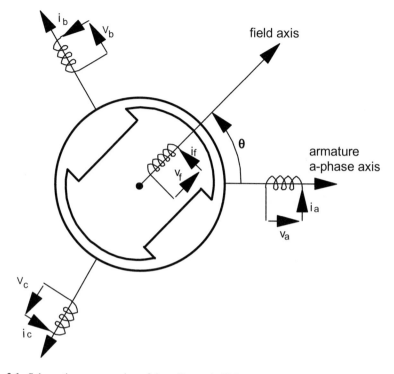

Fig. 2.1 Schematic representation of the salient pole SM

where V, I, and Φ are the armature voltage, current, and flux vectors which are expressed as follows:

$$V = \begin{bmatrix} v_a \\ v_b \\ v_c \end{bmatrix} \qquad I = \begin{bmatrix} i_a \\ i_b \\ i_c \end{bmatrix} \qquad \Phi = \begin{bmatrix} \phi_a \\ \phi_b \\ \phi_c \end{bmatrix} \qquad (2.2)$$

and where R is the resistance matrix, with:

$$R = \begin{bmatrix} r & 0 & 0 \\ 0 & r & 0 \\ 0 & 0 & r \end{bmatrix} \qquad (2.3)$$

where r is the resistance of an armature phase.
- Field:

$$v_f = r_f i_f + \frac{d\phi_f}{dt} \qquad (2.4)$$

$$
\text{where} \begin{cases} v_f & \text{the DC voltage feeding the field,} \\\\ i_f & \text{the DC current circulating in the field,} \\\\ \phi_f & \text{the field flux linkage,} \\\\ r_f & \text{the field resistance.} \end{cases}
$$

2.2.2 Magnetic Equation

Assuming that the magnetic circuit has a linear behaviour, the fluxes and the currents in the different circuits are linked by the following linear expressions:

$$
\begin{cases} \Phi = L_{ss}I + M_{sf}i_f \\\\ \phi_f = M_{fs}I + L_f i_f \end{cases} \tag{2.5}
$$

$$
\text{where} \begin{cases} L_{ss} & \text{square matrix of rank 3 including the armature self} - \text{and} \\\\ & \text{mutual inductances,} \\\\ M_{sf} & \text{single} - \text{column matrix including the mutual inductances between} \\\\ & \text{the armature phases and the field,} \\\\ M_{fs} & \text{transpose of matrix } M_{sf}, \\\\ L_f & \text{field self} - \text{inductance.} \end{cases}
$$

The variable saliency of the rotor makes the field MMF waveform far from being sinusoidal. Moreover, even though the armature winding is suitably arranged, its magnetic reaction still includes some harmonics. Consequently, the harmonic contents of the fluxes linking the windings (self and mutual) are rich in harmonics. Limiting their *Fourier* expansion to the fundamental terms enables a simple formulation of the inductance matrices.

Hence, L_{ss} could be expressed as the sum of two matrices, such that:

- L_{ss0} including constant inductances,

- L_{ss2} which includes inductances variables in terms of the electric angular position θ of the rotor with respect to the stator, as:

$$L_{ss} = L_{ss0} + L_{ss2} \qquad (2.6)$$

where:

$$L_{ss0} = \begin{bmatrix} L_{s0} & M_{s0} & M_{ss0} \\ M_{s0} & L_{s0} & M_{s0} \\ M_{s0} & M_{s0} & L_{s0} \end{bmatrix} \qquad (2.7)$$

with:

$$\begin{cases} L_{s0} = L_{ps0} + l_s \\ M_{s0} = -\frac{1}{2} L_{ps0} + m_s \end{cases} \qquad (2.8)$$

where $\begin{cases} L_{ps0} : \text{armature phase inductance due to the main flux,} \\ l_s : \quad \text{armature phase inductance due to the leakage flux,} \\ m_s : \quad \text{armature mutuelle inductance due to the leakage flux.} \end{cases}$

and where L_{ss2} is expressed in terms of θ as follows:

$$L_{ss2} = L_{s2} \begin{bmatrix} \cos 2\theta & \cos(2\theta - \frac{2\pi}{3}) & \cos(2\theta + \frac{2\pi}{3}) \\ \cos(2\theta - \frac{2\pi}{3}) & \cos(2\theta + \frac{2\pi}{3}) & \cos 2\theta \\ \cos(2\theta + \frac{2\pi}{3}) & \cos 2\theta & \cos(2\theta - \frac{2\pi}{3}) \end{bmatrix} \qquad (2.9)$$

Matrix M_{sf} is given by:

$$M_{sf} = M_0 \begin{bmatrix} \cos \theta \\ \cos(\theta - \frac{2\pi}{3}) \\ \cos(\theta + \frac{2\pi}{3}) \end{bmatrix} \qquad (2.10)$$

where M_0 is the maximum value of the mutual inductance between an armature phase and the field which is reached when their magnetic axis is aligned.

2.2.3 SM Park Model

2.2.3.1 Electrical Equation

The *Park* transform is systematically applied to the SM a-b-c model considering a dqo-frame linked to the rotor which is the same as the one linked to the rotating field as far as this latter and the rotor are moving synchronously. Of particular interest is the dqo-frame whose direct axis is aligned with the magnetic axis of the field which is commonly adopted in the literature.

Let us recall the relations linking the electrical angles θ, θ_s, and θ_r in such a dqo-frame:

$$\begin{cases} \theta_s = \theta \\ \theta_r = 0 \end{cases} \tag{2.11}$$

with:

$$\begin{cases} \dfrac{d\theta_s}{dt} = \omega_m \\ \dfrac{d\theta_r}{dt} = 0 \end{cases} \tag{2.12}$$

The application of the *Park* transform to the armature electrical Eq. (2.1) has led to:

$$\begin{cases} v_d = r i_d - \omega \phi_q + \dfrac{d\phi_d}{dt} \\ v_q = r i_q + \omega \phi_d + \dfrac{d\phi_q}{dt} \end{cases} \tag{2.13}$$

The application of the *Park* transform to the flux Eq. (2.5) has led to:

$$\begin{cases} \phi_d = L_d i_d + M i_f \\ \phi_q = L_q i_q \\ \phi_f = M i_d + L_f i_f \end{cases} \tag{2.14}$$

where L_d and L_q are the direct and quadrature inductances, respectively, with:

$$\begin{cases} L_d = L_{s0} - M_{s0} + \frac{3}{2}L_{s2} \\ L_q = L_{s0} - M_{s0} - \frac{3}{2}L_{s2} \end{cases} \tag{2.15}$$

and where:

$$M = \sqrt{\frac{3}{2}} M_{s0} \tag{2.16}$$

It is to be noted that, in smooth pole SM, the armature phase self- and mutual inductances are independent of the electrical position θ of the rotor with respect to the stator that yields:

$$L_{ss2} = \mathcal{O}_3 \tag{2.17}$$

Consequently, the direct and quadrature inductances turn to be:

$$L_d = L_q = L = L_{s0} - M_{s0} \tag{2.18}$$

2.2.3.2 Steady-State Phasor Representation

The qdo-frame is linked to the rotor. As far as the field is mounted on the rotor and is fed by a DC current (null frequency), then the armature variables expressed in the qdo-frame are continuous at steady state.

Under steady-state operation, Eq. (2.13) is reduced to:

$$\begin{cases} v_{d0} = ri_{d0} - \omega\phi_{q0} \\ v_{q0} = ri_{q0} + \omega\phi_{d0} \end{cases} \tag{2.19}$$

where subscript "0" indicates steady-state variables.

The substitution of ϕ_d and ϕ_q in Eq. (2.19), by their expressions given in Eq. (2.14), has led to:

$$\begin{cases} v_{d0} = ri_{d0} - L_q\omega i_{q0} \\ v_{q0} = ri_{q0} + L_d\omega i_{d0} + M\omega i_{f0} \end{cases} \tag{2.20}$$

Let us call:

$$\begin{cases} \overline{V} = v_{d0} + jv_{q0} \\ \overline{I} = i_{d0} + ji_{q0} \end{cases} \tag{2.21}$$

and let us define:

$$\begin{cases} \overline{\Phi} = Mi_{f0} \\ \overline{E} = j\omega\overline{\Phi} \end{cases} \tag{2.22}$$

where:

- $\overline{\Phi}$ is a vector aligned with the d-axis corresponding to the flux created by the field at steady-state operation. It should be noted that $\overline{\Phi}$ is different from ϕ_d introduced in Eq. (2.14). The equality of Φ and ϕ_d is satisfied at no-load generator operation.

- $\overline{E}$ is a vector corresponding to the steady-state back-EMF created by the field flux represented by vector $\overline{\Phi}$.

Let us define the armature current phasor as the following sum:

$$\overline{I} = \overline{I}_d + \overline{I}_q \tag{2.23}$$

where:

$$\begin{cases} \overline{I}_d = i_{d0} \\ \overline{I}_q = ji_{q0} \end{cases} \tag{2.24}$$

Then, the electrical equations (2.20) could be rewritten taking into consideration relations (2.22) and (2.24), which give the *Blondel* model of the SM, as:

$$\overline{V} = r\overline{I} + jX_d\overline{I}_d + jX_q\overline{I}_q + \overline{E} \tag{2.25}$$

where X_d and X_q are the direct and quadrature reactances, respectively, such that:

$$\begin{cases} X_d = L_d\omega \\ X_q = L_q\omega \end{cases} \tag{2.26}$$

The phasor diagram of Eq. (2.25) in the dq-frame is illustrated in Fig. 2.2 in the case where the armature magnetic reaction generates a direct flux which is opposite to the field one. Such scenario is known as "the flux weakening" which characterizes the high-speed operation of the SM. The phasor diagram, dealing with the case where both fluxes are additive, is shown in Fig. 2.3.

In the case of a smooth pole machine, $X_d = X_q = X = L\omega$ is called the synchronous reactance. Then, Eq. (2.25) is reduced to:

$$\overline{V} = r\overline{I} + jX\overline{I} + \overline{E} \tag{2.27}$$

yielding the *Behn-Eschenburg* model whose phasor representation is shown in Fig. 2.4.

2.2.3.3 Electromagnetic Torque Formulation

Let us consider, for instance, the motor operation of the salient pole SM under which it is fed by balanced three-phase voltages with an rms value V_{rms}, resulting in balanced currents and back-EMFs with rms values I_{rms} and E_{rms}, respectively.

Let us call T_{em1} the electromagnetic torque developed by a given phase:

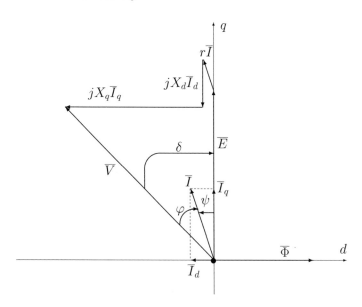

Fig. 2.2 Graphical representation in the case of the flux-weakening range ($\psi < 0$)

Fig. 2.3 Graphical representation in the case of the constant torque range ($\psi > 0$)

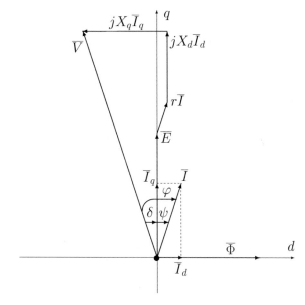

Fig. 2.4 Graphical
representation in the case of
a smooth pole SM operating
at $\psi = 0$

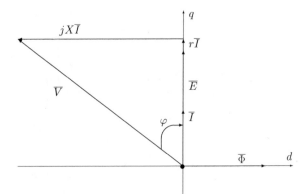

$$T_{em1} = \frac{P_{em1}}{\Omega_m} \tag{2.28}$$

where P_{em1} is the electromagnetic power developed by the phase, with:

$$P_{em1} = P_1 - r I_{rms}^2 \tag{2.29}$$

where P_1 is the power absorbed by the phase, with:

$$P_1 = V_{rms} I_{rms} \cos \varphi \tag{2.30}$$

Referring to Figs. 2.2 and 2.3, a relation between the angular shifts between phasors $\overline{V}, \overline{I}$, and $\overline{E}$ could be established:

$$\varphi = \delta + \psi \tag{2.31}$$

which yields:

$$V_{rms} I_{rms} \cos \varphi = V_{rms} I_{rms} (\cos \delta \cos \psi - \sin \delta \sin \psi) \tag{2.32}$$

The projection of Eq. (2.25) on the dqo-frame axis has led to the following relations:

$$\begin{cases} \|\overline{V}\| \sin \delta = X_q \|\overline{I}_q\| - r\|\overline{I}\| \sin \psi \\[2ex] \|\overline{V}\| \cos \delta = -X_d \|\overline{I}_d\| + r\|\overline{I}\| \cos \psi + \|\overline{E}\| \end{cases} \tag{2.33}$$

Following the substitution $\overline{I}_d$ and $\overline{I}_q$ by their expressions in terms of $\overline{I}$, Eq. (2.33) turns to be:

$$\begin{cases} \|\overline{V}\| \sin \delta = X_q \|\overline{I}\| \cos \psi - r \|\overline{I}\| \sin \psi \\ \\ \|\overline{V}\| \cos \delta = X_d \|\overline{I}\| \sin \psi + r \|\overline{I}\| \cos \psi + \|\overline{E}\| \end{cases} \quad (2.34)$$

Accounting for the relation between the direct and quadrature components and the rms value of an electromagnetic variable x:

$$\|\overline{X}\| = \sqrt{x_d^2 + x_q^2} = \sqrt{3} X_{rms} \quad (2.35)$$

and for relations (2.34), the equality (2.32) leads to:

$$V_{rms} I_{rms} \cos \varphi = I_{rms} \cos \psi \left(X_d I_{rms} \sin \psi + r I_{rms} \cos \psi + E_{rms} \right)$$

$$- I_{rms} \sin \psi \left(X_q I_{rms} \cos \psi - r I_{rms} \sin \psi \right) \quad (2.36)$$

The development of Eq. (2.36) yields:

$$V_{rms} I_{rms} \cos \varphi = E_{rms} I_{rms} \cos \psi + (X_d - X_q) I_{rms}^2 \cos \psi \sin \psi + r I_{rms}^2 \quad (2.37)$$

Taking into consideration expressions (2.28), (2.29), and (2.37), one of the electromagnetic torques is established:

$$T_{em} = 3 \left(\frac{E_{rms} I_{rms} \cos \psi + (X_d - X_q) I_{rms}^2 \cos \psi \sin \psi}{\Omega_m} \right) \quad (2.38)$$

which, accounting for relations (2.22) and (2.26), turns to be:

$$T_{em} = 3p \left(\Phi_{rms} I_{rms} \cos \psi + \frac{(L_d - L_q)}{2} I_{rms}^2 \sin 2\psi \right) \quad (2.39)$$

Accounting for the following:

$$\begin{cases} E_{rms} \quad = p\,\Omega_m\,\Phi_{rms} = p\,\Omega_m\,\dfrac{\|\overline{\Phi}\|}{\sqrt{3}} = p\,\Omega_m\,\dfrac{\phi}{\sqrt{3}} \\[4mm] I_{rms}\cos\psi = \dfrac{\|\overline{I}\|}{\sqrt{3}}\cos\psi = \dfrac{\|\overline{I_q}\|}{\sqrt{3}} \quad = \dfrac{i_q}{\sqrt{3}} \\[4mm] I_{rms}\sin\psi = \dfrac{\|\overline{I}\|}{\sqrt{3}}\sin\psi = -\dfrac{\|\overline{I_d}\|}{\sqrt{3}} \quad = \dfrac{i_d}{\sqrt{3}} \end{cases} \qquad (2.40)$$

the expression of the electromagnetic torque is finally reduced to:

$$T_{em} = p\,(\phi\,i_q + (L_d - L_q)\,i_d i_q) \qquad (2.41)$$

Referring to expression (2.39), the electromagnetic torque could be decomposed into two components, such that:

$$T_{em} = T_s + T_r \qquad (2.42)$$

where T_s and T_r are the synchronizing and reluctant torques, respectively, with:

$$\begin{cases} T_s = 3p\,\Phi_{rms}\,I_{rms}\,\cos\psi \\[3mm] T_r = 3p\,\dfrac{(L_d - L_q)}{2}\,I_{rms}^2\,\sin 2\psi \end{cases} \qquad (2.43)$$

Figures 2.5 and 2.6 show the variations of T_s, T_r, and T_{em} with respect to ψ (the so-called torque angle), in the cases of direct and reverse saliencies, respectively.

Referring to Fig. 2.5, it clearly appears that, in the case of a direct saliency ($L_d > L_q$), the reluctant torque contributes to the torque production capability of the salient pole SM in the range of the positive values of ψ. While for negative values of ψ, the reluctant torque behaves like a brake that affects the torque production capability. Nevertheless, such a drawback turns to be a requirement in order to achieve the flux weakening that enables the operation at high speeds.

From the analysis of expression (2.39), one can distinguish three values of the torque angle ψ, such that:

- $\psi = 0$ for which T_s is maximum and T_r is null. The salient pole SM behaves like a smooth pole one operating at maximum electromagnetic torque T_{em}^{max},
- $\psi = \frac{\pi}{4}$ for which T_r is maximum. In the case of a reversed saliency characterized by $L_d < L_q$, the torque angle yielding a maximum reluctant torque is $\psi = -\frac{\pi}{4}$,
- $\psi = \psi_M$ for which the electromagnetic torque is maximum.

In what follows, the analytical determination of ψ_M is carried out, considering the derivative of the expression of electromagnetic torque (2.39) with respect to ψ, as:

$$\frac{dT_{em}}{d\psi} = 3p\,I_{rms}(-\Phi_{rms}\sin\psi + (L_d - L_q)I_{rms}\cos 2\psi) \qquad (2.44)$$

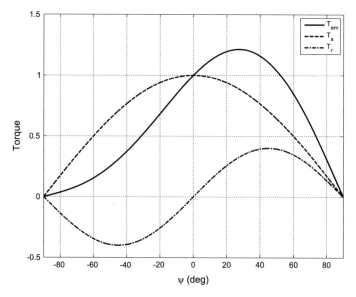

Fig. 2.5 Electromagnetic, synchronizing, and reluctant torques of a salient pole SM versus ψ in the case where $L_d > L_q$ (the so-called direct saliency)

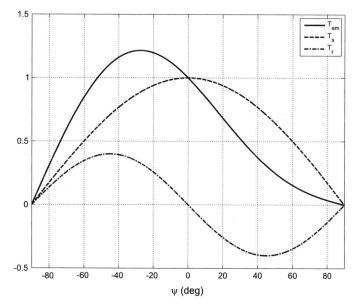

Fig. 2.6 Electromagnetic, synchronizing, and reluctant torques of a salient pole SM versus ψ in the case where $L_d < L_q$ (the so-called reverse saliency)

Thus, ψ_M fulfils the following equality:

$$k \sin \psi_M = \cos 2\psi_M \tag{2.45}$$

where:

$$k = \frac{\Phi_{rms}}{(L_d - L_q)I_{rms}} \tag{2.46}$$

The development of expression (2.45) leads to a second-order equation in terms of $\sin \psi_M$, as:

$$2 \sin^2 \psi_M + k \sin \psi_M - 1 = 0 \tag{2.47}$$

whose resolution yields:

- In the case of direct saliency $k > 0$ and $\psi_M > 0$ (as shown in Fig. 2.5):

$$\psi_{M1} = \arcsin \left(\frac{-k + \sqrt{k^2 + 8}}{4} \right) \tag{2.48}$$

- In the case of inverse saliency $k < 0$ and $\psi_M < 0$ (as shown in Fig. 2.6):

$$\psi_{M2} = \arcsin \left(\frac{-k - \sqrt{k^2 + 8}}{4} \right) = -\psi_{M1} \tag{2.49}$$

2.2.4 Operation at Maximum Torque

2.2.4.1 Case of Smooth Pole SMs

These machines could be excited by current or by PMs and could be equipped with distributed or concentrated windings in the armature. Referring to the previous paragraph, these machines develop a synchronizing torque, such that:

$$T_{em} = T_s = 3p \, \Phi_{rms} I_{rms} \cos \psi \tag{2.50}$$

The operation at maximum torque consists in feeding the armature circuits by currents with their initial phase equal to $\frac{\pi}{2}$ so that the armature current phasor, resulting from the application of the *Park* transform, is aligned with the back-EMF one. Doing so, the torque angle is kept null ($\psi = 0$), which gives:

$$T_{em}^{\psi=0} = 3p \, \Phi_{rms} I_{rms} \tag{2.51}$$

Furthermore, and referring to Fig. 2.4, the power factor is expressed as follows:

$$\cos \varphi = \frac{E_{rms} + r I_{rms}}{V_{rms}} \qquad (2.52)$$

Assuming that $r \ll X$ (condition commonly fulfilled except for low-speed operation), expression (2.52) turns to be:

$$\cos \varphi \simeq \frac{E_{rms}}{\sqrt{E_{rms}^2 + (X I_{rms})^2}} \qquad (2.53)$$

Giving the fact that $E_{rms} = \omega \Phi_{rms}$ and that $X = L\omega$, the power factor can be formulated as follows:

$$\cos \varphi \simeq \frac{1}{\sqrt{1 + \left(\frac{L I_{rms}}{\Phi_{rms}}\right)^2}} \qquad (2.54)$$

Consequently, for a given no-load flux (case of PM-excited SMs or case of DC-excited SMs operating at constant field current), the operation at maximum electromagnetic torque is characterized by a decrease in the power factor with the increase in the load torque. This requires the oversizing of the associated power electronic converter (the rating of the converter is higher than the machine one) which compromises the cost-effectiveness of the electric machine drive.

2.2.4.2 Case of Salient Pole SMs

Referring to Sect. 2.2.3.3, the maximum torque operation is achieved for a torque angle ψ_m, such that:

$$\psi_M = \arcsin\left(\frac{-k \pm \sqrt{k^2 + 8}}{4}\right) \qquad (2.55)$$

The substitution of k by its expression given in (2.46) in Eq. (2.55) gives:

$$\psi_M = \arcsin\left(\frac{-\Phi_{rms} + \sqrt{\Phi_{rms}^2 + 8(L_d - L_q)^2 I_{rms}^2}}{4(L_d - L_q) I_{rms}}\right) \qquad (2.56)$$

which is positive for a direct saliency ($L_d > L_q$) and negative for a reverse saliency. ($L_d < L_q$)

The corresponding synchronizing and reluctant torques are expressed as follows:

$$
\begin{cases}
T_s^{\psi=\psi_M} = 3p\ \Phi_{rms}\ I_{rms}\ \cos\arcsin\left(\dfrac{-\Phi_{rms} + \sqrt{\Phi_{rms}^2 + 8(L_d - L_q)^2 I_{rms}^2}}{4(L_d - L_q)I_{rms}}\right) \\[4mm]
\hspace{4cm}(2.57) \\[4mm]
T_r^{\psi=\psi_M} = 3p\ \dfrac{(L_d - L_q)}{2}\ I_{rms}^2\ \sin 2\arcsin\left(\dfrac{-\Phi_{rms} + \sqrt{\Phi_{rms}^2 + 8(L_d - L_q)^2 I_{rms}^2}}{4(L_d - L_q)I_{rms}}\right)
\end{cases}
$$

giving the fact that $\sin(2\psi_M) = 2\cos\psi_M \sin\psi_M$, one can establish the relation between $T_s^{\psi=\psi_M}$ and $T_r^{\psi=\psi_M}$ as follows:

$$
T_r^{\psi=\psi_M} = \left(\frac{\sqrt{1 + 8\left(\dfrac{(L_d - L_q)I_{rms}}{\Phi_{rms}}\right)^2} - 1}{4}\right) T_s^{\psi=\psi_M} \qquad (2.58)
$$

To sum up, for a given load torque, that is, to say for a given armature current, the operating point corresponding to the maximum electromagnetic torque is reached for a torque angle $\psi_M > 0$, yielding a lagging phasor $\overline{I}$ with respect to the back-EMF phasor $\overline{E}$ in the case of a direct saliency (see Fig. 2.3). While in the case of a reverse saliency, the operation at maximum electromagnetic torque is reached for a torque angle $\psi_M < 0$ that gives a leading armature current phasor $\overline{I}$ with respect to the back-EMF phasor $\overline{E}$ (see Fig. 2.2).

Giving the fact that the armature circuits are of R-L type, the current phasor $\overline{I}$ is usually lagging with respect to the armature voltage phasor $\overline{V}$. As far as this latter is usually leading with respect to the back-EMF phasor $\overline{E}$, one can conclude that the reverse saliency is more favourable to operate at a maximum electromagnetic torque with an improved power factor than the direct saliency [1, 2]. This statement is confirmed by Figs. 2.2 and 2.3.

2.2.5   Operation at Unity Power Factor

The operation at a unity power factor is characterized by $\varphi = 0$, that is, to say by the following relation:

$$
\delta = -\psi \qquad (2.59)
$$

with a negative torque angle ψ.

2.2.5.1 Case of Smooth Pole SMs

For the sake of simplicity, let us consider the case where $r \ll X$, and then:

$$\cos \psi = \frac{V_{rms}}{E_{rms}} \qquad (2.60)$$

where

$$E_{rms} = \omega \Phi_{rms} = \sqrt{V_{rms}^2 + (X I_{rms})^2} \qquad (2.61)$$

Rewriting expression (2.60), taking into account relation (2.61), gives:

$$\cos \psi = \sqrt{1 - \left(\frac{L I_{rms}}{\Phi_{rms}}\right)^2} \qquad (2.62)$$

Hence, under a unity power factor operation, the electromagnetic torque expression turns to be:

$$T_{em}^{\varphi=0} = 3p \, \Phi_{rms} I_{rms} \sqrt{1 - \left(\frac{L I_{rms}}{\Phi_{rms}}\right)^2} \qquad (2.63)$$

Referring to expression (2.63), one can notice a parabolic variation of $T_{em}^{\varphi=0}$ with respect to the rms value of the armature current I_{rms}. Indeed, it is to be noted that, beyond $I_{rms} = 0$, the electromagnetic torque $T_{em}^{\varphi=0}$ turns to be null for $I_{rms} = \frac{\Phi_{rms}}{L}$ which, accounting for expression (2.61), corresponds to a null armature voltage (short-circuited armature).

The derivative of the electromagnetic torque $T_{em}^{\varphi=0}$ with respect to I_{rms} has led to an armature current $I_{rms}^{T_{emax}}$, such that:

$$I_{rms}^{T_{emax}} = \frac{\Phi_{rms}}{\sqrt{2} \, L} \qquad (2.64)$$

The corresponding electromagnetic torque $T_{emax}^{\varphi=0}$ is expressed as follows:

$$T_{emax}^{\varphi=0} = \frac{3}{2} p \, \frac{\Phi_{rms}^2}{L} \qquad (2.65)$$

Referring to Sect. 2.2.4.1, it is to be noted that for a given armature current different from 0, the electromagnetic torque developed under a unity power factor operation is systematically lower than the one under a maximum torque one. Indeed, one can write the ratio:

$$\frac{T_{em}^{\varphi=0}}{T_{em}^{\psi=0}} = \sqrt{1 - \left(\frac{L I_{rms}}{\Phi_{rms}}\right)^2} < 1 \qquad (2.66)$$

Consequently, the usefulness of the armature current is better under a maximum torque operation than under a unity power factor one.

2.2.5.2 Case of Salient Pole SMs

Let us redraw Fig. 2.2 with $\psi = -\delta$, and let us neglect the voltage drop across r, and then one can establish the following expression:

$$\cos\psi = \frac{E_{rms} + X_d I_{rms} \sin\psi}{V_{rms}} \qquad (2.67)$$

Moreover, let us rewrite expression (2.37) accounting for $\varphi = 0$ and neglecting the voltage drop across r, as:

$$V_{rms} = E_{rms} \cos\psi + (X_d - X_q) I_{rms} \cos\psi \sin\psi \qquad (2.68)$$

Rewriting expression (2.67) taking into account equation (2.68) and substituting E_{rms} by $\omega\Phi_{rms}$ have led to:

$$\cos\psi = \frac{\Phi_{rms} + L_d I_{rms} \sin\psi}{\Phi_{rms} \cos\psi + (L_d - L_q) I_{rms} \cos\psi \sin\psi} \qquad (2.69)$$

whose development has given:

$$(1 - \cos^2\psi)\Phi_{rms} + L_d I_{rms} \sin\psi - (L_d - L_q) I_{rms} \cos^2\psi \sin\psi = 0 \qquad (2.70)$$

and finally:

$$(L_d - L_q) I_{rms} \sin\psi^2 + \Phi_{rms} \sin\psi + L_q I_{rms} = 0 \qquad (2.71)$$

The resolution of Eq. (2.71), taking into account the fact that $\psi < 0$ regardless of the saliency, has led to [1, 2]:

$$\psi_{\varphi=0} = \arcsin\left(\frac{-\Phi_{rms} + \sqrt{\Phi_{rms}^2 - 4(L_d - L_q) L_q I_{rms}^2}}{2(L_d - L_q) I_{rms}}\right) \qquad (2.72)$$

2.3 Flux-Weakening Operation of SMs

2.3.1 Similarity with the DC Machine

The expression of the speed of a DC motor with compensated armature magnetic reaction, regardless the type of excitation, could be expressed as follows:

$$\Omega_m = \frac{U_A - RI_A}{\frac{p}{a}\frac{N}{2\pi}\Phi_F} \tag{2.73}$$

where:
$$\begin{cases}
U_A & : \text{armature voltage,} \\
I_A & : \text{armture current,} \\
\Phi_F & : \text{field flux,} \\
R & : \text{total resistance through which the armature current is circulating,} \\
N & : \text{armature total actif conductors,} \\
p & : \text{pole pair,} \\
a & : \text{pairs of armature parallel circuits.}
\end{cases}$$

The motor develops an electromagnetic torque whose expression is as follows:

$$T_{em} = \frac{p}{a}\frac{N}{2\pi}\Phi_F\,I_A \tag{2.74}$$

For a given field flux Φ_F and for a given load torque, the armature current I_A turns to be constant. In order to vary the speed, the armature voltage U_A is varied using controlled rectifiers for average and high power drives and choppers for low power ones, yielding the so-called constant torque operation. The speed variation within this strategy is achieved when the armature maximum voltage is reached, leading to the so-called basic speed Ω_B, such that:

$$\Omega_B = \frac{U_{Amax} - RI_A}{\frac{p}{a}\frac{N}{2\pi}\Phi_F} \tag{2.75}$$

In order to have access to speeds higher than Ω_B, without exceeding the thermal limit of the armature circuit, one can simply reduce the field flux Φ_F at a constant armature voltage U_{Amax}, leading to the so-called flux-weakening operation. Keeping the armature current I_A constant, the flux weakening is characterized by an increase in the speed and a decrease in the electromagnetic torque, and therefore by a constant electromagnetic power P_{em}, as formulated in what follows:

$$P_{em} = T_{em}\Omega_m = (U_{Amax} - RI_A)I_A \tag{2.76}$$

2.3.2 SM Flux Weakening

2.3.2.1 Principle

Both smooth and salient pole SMs turn to be equivalent to a DC machine if the torque angle is kept null ($\psi = 0$). This yields an electromagnetic torque expression similar to the DC machine one, such that:

$$T_{em} = 3p\, \Phi_{rms}\, I_{rms} \tag{2.77}$$

In the manner of DC machines, SMs exhibit a capability to achieve a flux weakening through a reduction of the field flux. If such a flux reduction is easily carried out in DC-excited SMs with a decrease in the field current, it is not the case of the PM-excited SMs.

In fact, in contrary to DC machines, the armature magnetic reaction of SMs is never compensated. The flux-weakening operation of PM-excited SMs has been made feasible thanks to the armature magnetic reaction.

The basic idea consists in selecting a negative torque angle. The armature current phasor $\overline{I}$ is leading with respect to the back-EMF one $\overline{E}$, yielding a negative direct component of the armature current that produces a flux within the d-axis opposite to the one created by the PMs. This leads to an increase in the speed according to the formulation developed in the following paragraph [3].

2.3.2.2 Operation at Constant Torque

Let us consider a smooth pole PM-excited SM (for instance, the case of PMs mounted on the surface of the rotor) operating at a null torque angle ($\psi = 0$). Under high-speed operation, one can establish the following inequality:

$$X = L\omega = Lp\Omega_m \gg r \tag{2.78}$$

so that the voltage drop across r could be neglected.

Hence, the armature equation is reduced to:

$$\overline{V} = \overline{E} + jX\overline{I} \tag{2.79}$$

that gives:

$$V_{rms} = p\Omega_m \sqrt{\Phi_{rms}^2 + (L\, I_{rms})^2} \tag{2.80}$$

leading to:

$$\Omega_m = \frac{V_{rms}}{p\sqrt{\Phi_{rms}^2 + (L\,I_{rms})^2}} \tag{2.81}$$

Referring to expression (2.81), the speed variation should take into account the following limitations:

- The limitation dictated by the maximum armature current I_{max} that takes into account the limit of the armature circuit tolerable heating,
- The limitation dictated by the DC bus voltage at the inverter input, in other words the maximum armature voltage V_{max}.

Hence, increasing the speed at a constant electromagnetic torque (i.e. at constant armature current $I \leq I_{max}$) is feasible as long as the armature voltage fulfils the condition $V \leq V_{max}$.

The speed corresponding to $V = V_{max}$, the so-called base speed Ω_b, is expressed as follows:

$$\Omega_b = \frac{V_{max}}{p\sqrt{\Phi_{rms}^2 + (L\,I_{rms})^2}} \tag{2.82}$$

Figure 2.7 illustrates the principle of the speed variation by controlling the armature voltage, starting from a speed Ω_1 until Ω_b. It should be noted that the phasors $\overline{E} = j\omega\Phi$ and $jX\overline{I} = jL\omega\overline{I}_1$ are proportional to the speed. Consequently, the speed increase from Ω_1 to Ω_b is achieved at constant angle δ.

Reaching the speed Ω_b, a crucial question arises: How could higher speeds be accessed?

To do so, two approaches could be considered which are described in the following paragraphs.

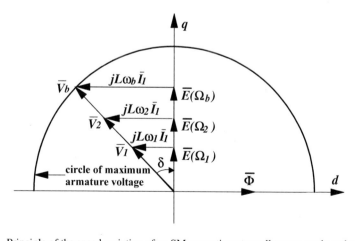

Fig. 2.7 Principle of the speed variation of an SM, operating at a null torque angle and a constant armature current, based on the variation of the armature voltage, leading to a speed range varying from 0 to Ω_b

2.3.2.3 Speed Variation at Null Torque Angle

The first approach consists in keeping the torque angle ψ equal to zero and the armature voltage at its maximum value, as illustrated in Fig. 2.8.

Giving the fact that the operating points are located on the circle of maximum armature voltage, these are characterized by speeds expressed as follows:

$$\Omega_m = \frac{V_{max}}{p \sqrt{\Phi_{rms}^2 + (L\, I_{rms})^2}} \tag{2.83}$$

The high-speed operation of SMs is commonly characterized by the so-called flux ratio $\mathcal{R}_\varphi$, defined as:

$$\mathcal{R}_\varphi = \frac{L\, I_{rms}}{\Phi_{rms}} \tag{2.84}$$

which depends on the machine design (Φ_{rms} and L) on one hand and the operating point (I_{rms}) on the other hand. It is the ratio of the flux corresponding to the armature magnetic reaction to the excitation flux.

Accounting for the expression of $\mathcal{R}_\varphi$, one of the speeds given in (2.83) turns to be:

$$\Omega_m = \frac{V_{max}}{P \Phi_{rms} \sqrt{1 + \mathcal{R}_\varphi^2}} \tag{2.85}$$

In order to reach speeds higher than Ω_b, a significant decrease in the armature current has to be carried out. Indeed, referring to Fig. 2.8, the modulus of phasor $j L \omega_2 \overline{I}_2$ is lower than one of the phasors $j L \omega_b \overline{I}_1$. Moreover, $\omega_2 > \omega_b$, which leads to a remarkable decrease in the armature current and consequently in the electromagnetic torque. In spite of the speed increase, the electromagnetic power falls sharply to zero at a maximum speed Ω_{max1}, such that:

$$\Omega_{max1} = \frac{V_{max}}{p\, \Phi_{rms}} \tag{2.86}$$

To sum up, it comes out that an increase in the speed of PM-excited SMs according to the above-described approach does not meet the similitude of the targeted constant electromagnetic power operation of the DC machines.

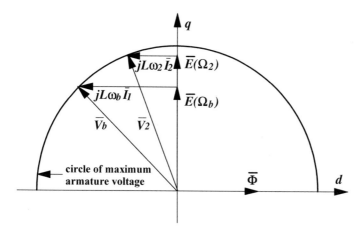

Fig. 2.8 Principle of the speed variation of an SM, operating at a null torque angle and maximum armature voltage, based on the variation of the armature current, leading to a speed range varying from Ω_b to Ω_{max1}

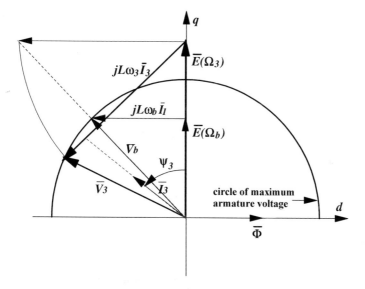

Fig. 2.9 Principle of the speed variation of an SM operating at a constant armature current and maximum armature voltage, based on the variation of the torque angle, leading to a speed range varying from Ω_b to Ω_{max2} if $\mathcal{R}_\varphi > 1$, or to Ω_{max3} if $\mathcal{R}_\varphi < 1$, or to an infinite speed if $\mathcal{R}_\varphi = 1$

2.3.2.4 Speed Variation at Constant Armature Current

The second approach consists in maintaining constant armature current I_{rms} and voltage $V_{rms} = V_{max}$, while reducing the torque angle ψ (from 0 to negative values) in such a way that $\overline{I}$ is constantly leading with respect to $\overline{E}$, as shown in Fig. 2.9.

The graphical representation considers, in a first step, that the armature voltage could exceed its maximum value so that the graphical representation can be done in a similar way as in the case of the constant torque operation. Then, the extremity of the obtained phasor $jX\overline{I}$ is moved along a circle that has a radius XI_{rms} and a centre the extremity of the phasor $\overline{E}$, until the interception of the circle of maximum armature voltage. The resulting operating point is characterized by the following armature voltage equation:

$$\overline{V}_3 = \overline{E}(\Omega_3) + jL\omega_3\overline{I}_3 \tag{2.87}$$

Then, the application of the generalized form of the *Pythagoras* theorem is expressed as follows:

$$\|\overline{V}_3\|^2 = \|jL\omega_3\overline{I}_3\|^2 - 2\|\overline{E}(\Omega_3)\|\|jL\omega_3\overline{I}_3\|\cos\gamma + \|\overline{E}(\Omega_3)\|^2 \tag{2.88}$$

where:

$$\gamma = \widehat{(\overline{E}(\Omega_3), jL\omega_3\overline{I}_3)} \tag{2.89}$$

Angles ψ and γ are linked by the following relation:

$$\gamma = \psi_3 + \frac{\pi}{2} \quad \text{with} \quad \psi_3 < 0 \tag{2.90}$$

Thus, Eq. (2.88) turns to be:

$$V_{rms}^2 = (L\omega_3 I_{rms})^2 + 2L\omega_3^2 I_{rms}\Phi_{rsm}\sin\psi_3 + (\omega_3\Phi_{rsm})^2 \tag{2.91}$$

which leads to:

$$\Omega_3 = \frac{V_{max}}{p\sqrt{(LI_{rms})^2 + 2LI_{rms}\Phi_{rms}\sin\psi_3 + \Phi_{rms}^2}} \tag{2.92}$$

which can be rewritten in terms of the flux ratio, as:

$$\Omega_3 = \frac{V_{max}}{p\Phi_{rms}\sqrt{\mathcal{R}_\varphi^2 + 2\mathcal{R}_\varphi\sin\psi_3 + 1}} \tag{2.93}$$

Three scenarios could be distinguished, namely:

- $\mathcal{R}_\varphi > 1$, in this case, a reduction of the torque angle ψ from 0 to $-\frac{\pi}{2}$ leads to a displacement of the operating point along the circle of maximum armature voltage until reaching the point $(0, -V_{max})$ for the maximum speed:

$$\Omega_{max2} = \frac{V_{max}}{p\,\Phi_{rms}(\mathcal{R}_\varphi - 1)} \tag{2.94}$$

- $\mathcal{R}_\varphi < 1$, in this case, a reduction of the torque angle ψ from 0 to $-\frac{\pi}{2}$ leads to a displacement of the operating point along the circle of maximum armature voltage until reaching the point $(0, V_{max})$ for the maximum speed:

$$\Omega_{max3} = \frac{V_{max}}{p\,\Phi_{rms}(1 - \mathcal{R}_\varphi)} \tag{2.95}$$

- $\mathcal{R}_\varphi = 1$, in this case, a reduction of ψ from 0 to $-\frac{\pi}{2}$ leads to an infinite speed range.

2.3.2.5 Case of Low Power Drives

In the case of low power drives, the smooth pole surface-mounted PM synchronous motor has an armature resistance r which could not be neglected in a wide speed range.

Operation in the Constant Torque Range The phasor diagram characterizing the steady-state operation of the synchronous motor in the constant torque region turns to be the one shown in Fig. 2.4. The application of the *Pythagoras* theorem yields:

$$V_{rms}^2 = (\omega\Phi_{rms} + r\,I_{rms})^2 + (L\omega I_{rms})^2 \tag{2.96}$$

that gives:

$$\left(\Phi_{rms}^2 + (LI_{rms})^2\right)\omega^2 + 2r\Phi_{rms}I_{rms}\omega + (r\,I_{rms})^2 - V_{rms}^2 = 0 \tag{2.97}$$

Then, the expression of the base speed turns to be:

$$\Omega_B = \frac{\sqrt{(r\Phi_{rms}I_{rms})^2 + \left(\Phi_{rms}^2 + (LI_{rms})^2\right)\left(V_{max}^2 - (r\,I_{rms})^2\right)} - r\Phi_{rms}I_{rms}}{p\left(\Phi_{rms}^2 + (LI_{rms})^2\right)} \tag{2.98}$$

which can be expressed in terms of the flux ratio $\mathcal{R}_\varphi$ as follows:

$$\Omega_B = \frac{\sqrt{(r\,I_{rms})^2 + \left(1 + \mathcal{R}_\varphi^2\right)\left(V_{max}^2 - (r\,I_{rms})^2\right)} - r\,I_{rms}}{p\,\Phi_{rms}\left(1 + \mathcal{R}_\varphi^2\right)} \tag{2.99}$$

or as:

$$\Omega_B = \frac{r\,I_{rms}}{p\,\Phi_{rms}}\left(\frac{\sqrt{1 + \left(1 + \mathcal{R}_\varphi^2\right)\left(\left(\frac{V_{max}}{r\,I_{rms}}\right)^2 - 1\right)} - 1}{\left(1 + \mathcal{R}_\varphi^2\right)}\right) \tag{2.100}$$

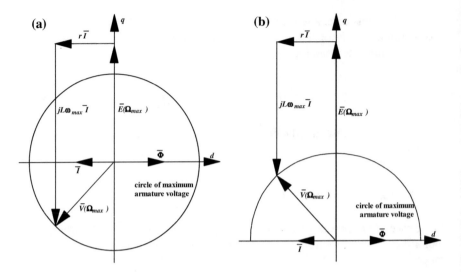

Fig. 2.10 Phasor diagram of the PM synchronous machine, with the armature phase resistance accounted for, under maximum speed operation achieved by a maximum armature voltage and a torque angle $\psi = -\frac{\pi}{2}$. **Legend: a** case where $\mathcal{R}_\varphi > 1$ and **b** case where $\mathcal{R}_\varphi < 1$

Operation in the Flux-Weakening Range Adopting a speed variation technique based on the same approach treated in Sect. 2.3.2.4 leads to the three following scenarios:

- $\mathcal{R}_\varphi > 1$, in this case, a reduction of the torque angle ψ from 0 to $-\frac{\pi}{2}$ leads to a displacement of the operating point along the circle of maximum armature voltage until reaching the point shown in Fig. 2.10a, for which the speed takes a maximum value Ω_{max2r}, with:

$$\Omega_{max2r} = \frac{\sqrt{V_{max}^2 - (r I_{rms})^2}}{p \, \Phi_{rms} (\mathcal{R}_\varphi - 1)} \tag{2.101}$$

- $\mathcal{R}_\varphi < 1$, in this case, a reduction of the torque angle ψ from 0 to $-\frac{\pi}{2}$ leads to a displacement of the operating point along the circle of maximum armature voltage until reaching the point shown in Fig. 2.10b, for which the speed takes a maximum value Ω_{max3r}, with:

$$\Omega_{max3} = \frac{\sqrt{V_{max}^2 - (r I_{rms})^2}}{p \, \Phi_{rms} (1 - \mathcal{R}_\varphi)} \tag{2.102}$$

- $\mathcal{R}_\varphi = 1$, in this case, a reduction of ψ from 0 to $-\frac{\pi}{2}$ leads to an infinite speed range.

2.3.2.6 Case of Salient Pole SMs

Let us consider, for instance, the case of a salient pole SM with a positive saliency. Its steady-state operation is represented by the phasor diagram shown in Fig. 2.3.

Operation in the Constant Torque Range Different strategy could be adopted under variable speed operation, such as:

- Operation at maximum synchronizing torque ($\psi = 0$),
- Operation at maximum torque ($\psi = \psi_m$).

Operation at Maximum Synchronizing Torque

The salient pole SM operation turns to be similar to a smooth pole one. Its phase diagram turns to be similar to the one shown in Fig. 2.4 where the reactance X should be replaced by the quadrature one X_q. Consequently, one can easily deduce the expression of the base speed from the one given in Eq. (2.100), as:

$$\Omega_B = \frac{r\,I_{rms}}{p\,\Phi_{rms}} \left(\frac{\sqrt{1 + \left(1 + \mathcal{R}_{\varphi q}^2\right)\left(\left(\frac{V_{max}}{r\,I_{rms}}\right)^2 - 1\right)} - 1}{\left(1 + \mathcal{R}_{\varphi q}^2\right)} \right) \qquad (2.103)$$

where:

$$\mathcal{R}_{\varphi q} = \frac{L_q\,I_{rms}}{\Phi_{rms}} \qquad (2.104)$$

Neglecting the armature phase resistance, the expression of the base speed (2.103) is reduced as:

$$\Omega_B = \frac{V_{max}}{p\Phi_{rms}\,\sqrt{\left(1 + \mathcal{R}_{\varphi q}^2\right)}} \qquad (2.105)$$

Operation at Maximum Torque

The application of the *Pythagoras* theorem to the phasor diagram of Fig. 2.3 yields the following equation:

$$(\Phi_{rms}\omega + r I_{rms}\cos\psi_m + L_d\omega I_{rms}\sin\psi_m)^2 + (L_q\omega I_{rms}\cos\psi_m - r I_{rms}\sin\psi_m)^2 = V_{rms}^2 \qquad (2.106)$$

whose development leads to:

$$((\Phi_{rms} + L_d I_{rms}\sin\psi_m)^2 + (L_q I_{rms}\cos\psi_m)^2)\,\omega^2 +$$

$$2r I_{rms}\cos\psi_m(\Phi_{rms} + I_{rms}\sin\psi_m(L_d - L_q))\,\omega + (r I_{rms})^2 - V_{rms}^2 = 0 \qquad (2.107)$$

The resolution of Eq. (2.107), considering the maximum value of the armature voltage, leads to the following expression of the base speed:

$$\Omega_B = \frac{r I_{rms} \cos \psi_m \left(\sin \psi_m (L_q - L_d) - \Phi_{rms} \right) + \sqrt{\Delta}}{p \left((\Phi_{rms} + L_d I_{rms} \sin \psi_m)^2 + (L_q I_{rms} \cos \psi_m)^2 \right)} \qquad (2.108)$$

where:

$$\Delta = \left(r I_{rms} \cos \psi_m \left(\sin \psi_m (L_q - L_d) - \Phi_{rms} \right) \right)^2$$

$$+ ((\Phi_{rms} + L_d I_{rms} \sin \psi_m)^2 + (L_q I_{rms} \cos \psi_m)^2)(V_{max}^2 - (r I_{rms})^2) \qquad (2.109)$$

Neglecting the armature phase resistance, the expression of the base speed (2.108) is reduced as:

$$\Omega_B = \frac{V_{max}}{p \Phi_{rms} \sqrt{(1 + \mathcal{R}_{\varphi d} \sin \psi_m)^2 + (\mathcal{R}_{\varphi q} \cos \psi_m)^2}} \qquad (2.110)$$

where:

$$\mathcal{R}_{\varphi d} = \frac{L_d I_{rms}}{\Phi_{rms}} \qquad (2.111)$$

Operation in the Flux-Weakening Range Reaching Ω_B, one can keep increasing the speed considering an approach similar to the one treated in Sect. 2.3.2.4. This leads to the three following scenarios:

- $\mathcal{R}_{\varphi d} > 1$, in this case, a reduction of the torque angle ψ from 0 or ψ_m to $-\frac{\pi}{2}$ leads to a displacement of the operating point along the circle of maximum armature voltage until reaching a maximum speed Ω_{max2sp}. This latter is characterized by a phasor diagram similar to the one shown in Fig. 2.10a where L is substituted by L_d and is expressed as follows:

$$\Omega_{max2sp} = \frac{\sqrt{V_{max}^2 - (r I_{rms})^2}}{p \, \Phi_{rms} (\mathcal{R}_{\varphi d} - 1)} \qquad (2.112)$$

- $\mathcal{R}_{\varphi d} < 1$, in this case, a reduction of the torque angle ψ from 0 or ψ_m to $-\frac{\pi}{2}$ leads to a displacement of the operating point along the circle of maximum armature voltage until reaching a maximum speed Ω_{max3sp}. This latter is characterized by a phasor diagram similar to the one shown in Fig. 2.10b where L is substituted by L_d and is expressed as follows:

$$\Omega_{max3sp} = \frac{\sqrt{V_{max}^2 - (r I_{rms})^2}}{p \, \Phi_{rms} (1 - \mathcal{R}_{\varphi d})} \qquad (2.113)$$

- $\mathcal{R}_{\varphi d} = 1$, in this case, a reduction of ψ from 0 to $-\frac{\pi}{2}$ leads to an infinite speed range.

2.4 Case Study

Figure 2.11 illustrates the block diagram of a series/parallel hybrid propulsion system. It includes an electric drive unit and a thermal power one. This latter is built around an internal combustion engine (ICE). Made up of a planetary gear, the power split device enables the control of the complementary ratios of the ICE-produced mechanical power applied to the traction wheels through the gearbox and to the generator shaft.

The electric drive unit is built around a PM-excited synchronous motor. It is equipped with a three-phase distributed winding in the armature and surface-mounted PMs in the rotor, with a pole pair $p = 2$. Moreover, the voltage drop across the armature resistance is supposed to be negligible in the whole speed range.

When the motor absorbs its rated current I_{rms}^r, it exhibits:

- A constant torque region of [0 1200] rpm, developing an electromagnetic torque of 400 Nm,
- A flux-weakening range reaching 6000 rpm at a null electromagnetic power, achieved following a reduction of the torque angle ψ,
- A flux ratio $\mathcal{R}_\varphi^r$, defined as:

$$\mathcal{R}_\varphi^r = \frac{L\, I_{rms}^r}{\Phi_{rms}} < 1 \tag{2.114}$$

The motor is fed by a battery pack through an IGBT inverter of 50 kW. The battery pack is composed of 168 NiMH-made storage cells providing a total of 201.6 V DC. A boost chopper enables the increase in such a voltage to reach a DC bus voltage $U_{DC} = 500$ V at the inverter input.

Starting from the base speed, the armature voltage is kept at its maximum value $V_{max} = \frac{\sqrt{2}}{\pi} U_{DC}$; the voltage drops across the inverter switches as well as the dead

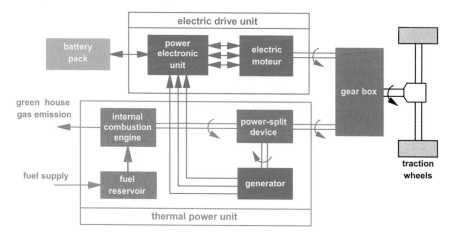

Fig. 2.11 Block diagram of a series/parallel hybrid propulsion system

times, considered in order to avoid the simultaneous conduction of the switches of the same leg, are omitted.

2.4.1 Part 1: Current Motor

Question Find out the flux ratio $\mathcal{R}_\varphi^r$.

Answer The base speed Ω_B is expressed in terms of the rated flux ratio $\mathcal{R}_\varphi^r$ as follows:

$$\Omega_B = \frac{V_{max}}{p\Phi_{rms}\sqrt{1 + \left(\mathcal{R}_\varphi^r\right)^2}} \tag{2.115}$$

The maximum speed Ω_{max} is expressed in terms of the rated flux ratio $\mathcal{R}_\varphi^r$ as follows:

$$\Omega_{max} = \frac{V_{max}}{p\,\Phi_{rms}\left(1 - \mathcal{R}_\varphi^r\right)} \tag{2.116}$$

The expression $\left(\frac{\Omega_{max}}{\Omega_B}\right)^2$ gives:

$$\left(\frac{\Omega_{max}}{\Omega_B}\right)^2 = \frac{1 + \left(\mathcal{R}_\varphi^r\right)^2}{\left(1 - \mathcal{R}_\varphi^r\right)^2} \tag{2.117}$$

whose development leads to:

$$\left(\left(\frac{\Omega_{max}}{\Omega_B}\right)^2 - 1\right)\left(\mathcal{R}_\varphi^r\right)^2 - 2\left(\frac{\Omega_{max}}{\Omega_B}\right)^2 \mathcal{R}_\varphi^r + \left(\frac{\Omega_{max}}{\Omega_B}\right)^2 - 1 = 0 \tag{2.118}$$

The numerical application yields:

$$24\left(\mathcal{R}_\varphi^r\right)^2 - 50\mathcal{R}_\varphi^r + 24 = 0 \tag{2.119}$$

Giving the fact that $\mathcal{R}_\varphi^r$ is lower than unity, the solution is then:

$$\boxed{\mathcal{R}_\varphi^r = 0.75}$$

Question Find out the PM flux Φ_{rms}.

Answer Φ_{rms} is expressed in terms of the maximum speed as follows:

$$\Phi_{rms} = \frac{V_{max}}{p\Omega_{max}\left(1 - \mathcal{R}_\varphi^r\right)} \tag{2.120}$$

which gives:

$$\boxed{\Phi_{rms} \simeq 716\,\text{mWb}}$$

Question Find out the rated current I_{rms}^r.
Answer In the constant torque region and for the rated armature current, the electro-magnetic torque T_{em}^r is 400 Nm. This later could be expressed as follows:

$$T_{em}^r = 3p\Phi_{rms}I_{rms}^r \tag{2.121}$$

Then:

$$\boxed{I_{rms}^r \simeq 93\,\text{A}}$$

Question Find out the power factor at the operating point corresponding to the base speed.
Answer Neglecting the armature resistance, the electromagnetic power P_{em} turns to be equal to the absorbed one which yields, for the base speed, the following relation:

$$P_{em}^r = T_{em}^r \Omega_B = 3V_{max}I_{rms}^r \cos\varphi_B \tag{2.122}$$

Then:

$$\cos\varphi_B = \frac{T_{em}^r \Omega_B}{3V_{max}I_{rms}^r} \tag{2.123}$$

which, accounting for the expression of electromagnetic torque given in Eq. (2.121), leads to:

$$\cos\varphi_B = \frac{p\Phi_{rms}\Omega_B}{V_{max}} \tag{2.124}$$

The numerical application gives:

$$\boxed{\cos\varphi_B = 0.8}$$

Question Find out the trajectories described by the extremities of the armature voltage phasor $\overline{V}$ and current one $\overline{I}$ in the dq-plane, considering the total speed range.
Answer see Fig. 2.12.

2.4.2 Projected Motor

For the sake of an extension of the flux-weakening range, one could investigate the solution consisting in increasing the inductance L through the substitution of the armature-distributed winding by a concentrated fractional-slot one.

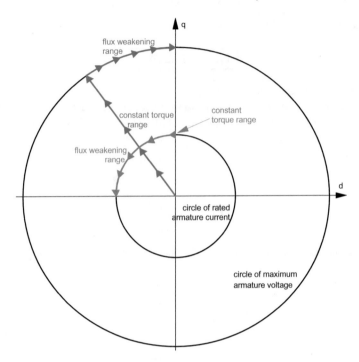

Fig. 2.12 Trajectories described by the extremities of the armature voltage phasor $\overline{V}$ (in blue) and current one $\overline{I}$ (in red) in the dq-plane considering the total speed range

Question Assuming an infinite extension of the flux-weakening range, find out: the inductance ratio $\dfrac{L_\infty}{L}$.

Answer The ratio $\dfrac{\mathcal{R}_F^\infty}{\mathcal{R}_F^r}$ is expressed as follows:

$$\frac{\mathcal{R}_F^\infty}{\mathcal{R}_F^r} = \left(\frac{\dfrac{L_\infty I_{rms}^r}{\Phi_{rms}}}{\dfrac{L I_{rms}^r}{\Phi_{rms}}} \right) = \frac{1}{0.75} \qquad (2.125)$$

Thus:

$$\boxed{\frac{L_\infty}{L} \simeq 133\%}$$

that is, to say a 33% increase in L which is feasible following the substitution of the armature-distributed winding by a concentrated one.

Question Find out the base speed $\Omega_{B\infty}$ (in rpm).

Answer Under a unity flux ratio, the expression of the base speed turns to be:

$$\Omega_{B\infty} = \frac{V_{max}}{p\Phi_{rms}\sqrt{2}} \qquad (2.126)$$

which gives:

$$\boxed{\Omega_{B\infty} \simeq 1061\,\text{rpm}}$$

Question Find out the electromagnetic power $P_{em\infty}$ in the flux-weakening range.
Answer Under a unity flux ratio, the flux-weakening range turns to be infinite and the electromagnetic power $P_{em\infty}$ turns to be almost constant and tends to its maximum value:

$$P_{em\infty} = 3I_{rms}V_{max} \qquad (2.127)$$

which gives:

$$\boxed{P_{em\infty} \simeq 62.83\,\text{kW}}$$

Question Find out the power factor at the operating point corresponding to the base speed.
Answer In the manner of the current motor, the power factor of the projected one is expressed as follows:

$$\cos\varphi_\infty = \frac{p\Phi_{rms}\Omega_{B\infty}}{V_{max}} \qquad (2.128)$$

which, accounting for the expression of $\Omega_{B\infty}$ given by Eq. (2.126), leads to:

$$\cos\varphi_\infty = \frac{1}{\sqrt{2}} \qquad (2.129)$$

and then:

$$\boxed{\cos\varphi_\infty = 0.707}$$

To sum up, one can notice that an infinite flux-weakening range is achieved with:

- A decrease in the constant torque region with a base speed of 1061 rpm instead of 1200 rpm,
- A decrease in the power factor from 0.8 to 0.707, at the base speed.

2.4.3 Comparative Study

Question Find out the characteristics giving T_{em} and P_{em} in terms of Ω_m (in rpm) assuming an infinite extension of the flux-weakening range. For the sake of comparison, $T_{em}(\Omega_m)$ and $P_{em}(\Omega_m)$ in the case of the current motor should be also plotted.
Answer

- Constant Torque Range

  Currentmotor

  for $N_m \in [0 \quad 1200]$rpm

$$\begin{cases} T_{em}^r = 400 \\[2mm] P_{em}^r = 400\Omega_m \end{cases} \tag{2.130}$$

  Projectedmotor

  for $N_m \in [0 \quad 1061]$rpm

$$\begin{cases} T_{em}^\infty = 400 \\[2mm] P_{em}^\infty = 400\Omega_m \end{cases} \tag{2.131}$$

- Flux-Weakening Range

  The speed is expressed as:

$$\Omega_m = \frac{V_{max}}{p\Phi_{rms}\sqrt{\mathcal{R}_\varphi^2 + 2\mathcal{R}_\varphi \sin\psi + 1}} \tag{2.132}$$

which gives:

$$\sin\psi = \frac{\left(\dfrac{V_{max}}{p\Phi_{rms}\Omega_m}\right)^2 - (\mathcal{R}_\varphi^2 + 1)}{2\mathcal{R}_\varphi} \tag{2.133}$$

The electromagnetic torque T_{em} is then expressed in terms of the speed and the flux ratio as follows:

$$T_{em} = 3p\Phi_{rms}I_{rms}\cos\left(\arcsin\left(\frac{\left(\dfrac{V_{max}}{p\Phi_{rms}\Omega_m}\right)^2 - (\mathcal{R}_\varphi^2 + 1)}{2\mathcal{R}_\varphi}\right)\right) \tag{2.134}$$

so does the expression of the electromagnetic power P_{em}:

$$P_{em} = T_{em}\Omega_m = 3p\Phi_{rms}I_{rms}\cos\left(\arcsin\left(\frac{\left(\dfrac{V_{max}}{p\Phi_{rms}\Omega_m}\right)^2 - (\mathcal{R}_\varphi^2 + 1)}{2\mathcal{R}_\varphi}\right)\right)\Omega_m \tag{2.135}$$

Currentmotor

for $N_m \in [1200 \quad 6000]$rpm

The expressions of the electromagnetic torque T_{em}^r and power P_{em}^r are deduced from Eqs. (2.134) and (2.135), respectively, where $I_{rms} = I_{rms}^r$ and $\mathcal{R}_\varphi = \mathcal{R}_\varphi^r$ as follows:

$$\begin{cases} T_{em}^r = 3p\Phi_{rms}I_{rms}^r \cos\left(\arcsin\left(\frac{\left(\frac{V_{max}}{p\Phi_{rms}\Omega_m}\right)^2 - \left(\left(\mathcal{R}_\varphi^r\right)^2 + 1\right)}{2\mathcal{R}_\varphi^r}\right)\right) \\[6ex] P_{em}^r = 3p\Phi_{rms}I_{rms}^r \cos\left(\arcsin\left(\frac{\left(\frac{V_{max}}{p\Phi_{rms}\Omega_m}\right)^2 - \left(\left(\mathcal{R}_\varphi^r\right)^2 + 1\right)}{2\mathcal{R}_\varphi^r}\right)\right) \Omega_m \end{cases}$$

(2.136)

Projectedmotor

for $N_m \in [1061 \quad \infty]$rpm

The expressions of the electromagnetic torque T_{em}^∞ and power P_{em}^∞ are deduced from Eqs. (2.134) and (2.135), respectively, where $I_{rms} = I_{rms}^r$ and $\mathcal{R}_\varphi = 1$ as follows:

$$\begin{cases} T_{em}^\infty = 3p\Phi_{rms}I_{rms}^r \cos\left(\arcsin\left(\frac{\left(\frac{V_{max}}{p\Phi_{rms}\Omega_m}\right)^2 - 2}{2}\right)\right) \\[6ex] P_{em}^\infty = 3p\Phi_{rms}I_{rms}^r \cos\left(\arcsin\left(\frac{\left(\frac{V_{max}}{p\Phi_{rms}\Omega_m}\right)^2 - 2}{2}\right)\right) \Omega_m \end{cases}$$

(2.137)

The derived torque–speed $T_{em}(\Omega_m)$ and power–speed $P_{em}(\Omega_m)$ characteristics have been plotted, for both current and projected motors. They are illustrated in Fig. 2.13. One can clearly notice the great similarity of the characteristics of the projected motor and the DC motor ones. Such a statement has been confirmed and experimentally validated in [4]. In spite of this superiority, it should be underlined that fractional-slot PM synchronous machines suffer from a dense harmonic content of the air gap flux density which induces high eddy current loss in the PMs that could lead to their demagnetization [5, 6]. Hopefully, this drawback has been eradicated thanks to the PM segmentation [7].

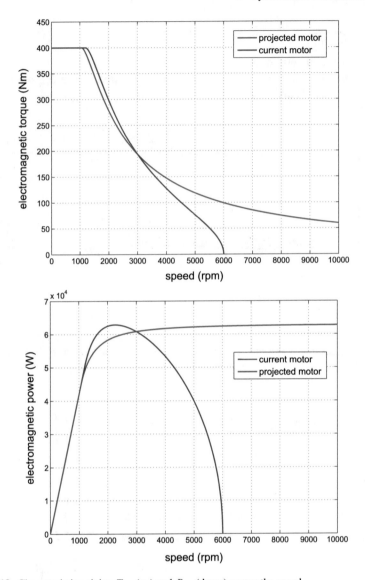

Fig. 2.13 Characteristics giving T_{em} (up) and P_{em} (down) versus the speed

References

1. M. Lajoie-Mazenc, Ph. Viarouge, *Power Supply of Synchronous Machines* (in French) (Technique de l'Ingenieur, D3630, Paris, 1991)
2. G. Grellet, G. Clerc, *Electrical Acuators: Principles/Models/Control* (in French) (Eyrolles, Paris, 1997). ISBN: 2-212-09352-7

3. B. Multon, J. Lucidarme, L. Prévond, Analysis of the possibilities of the flux weakening of permanent magnet machines (in French). J. Phys. III **5**, 623–640 (1995)
4. A.M. El-Refaie, T.M. Jahns, P.J. McCleer, J.W. McKeever, Experimental verification of optimal flux weakening in surface PM machines using concentrated windings. IEEE Trans. Ind. Appl. **42**(2), 443–53 (2006)
5. D. Ishak, Z.Q. Zhu, D. Howe, Eddy-current loss in the rotor magnets of permanent-magnet brushless machines having a fractional number of slots per pole. IEEE Trans. Magn. **41**(9), 2462–2469 (2005)
6. K. Yamazaki, Y. Fukushima, M. Sato, Loss analysis of permanent-magnet motors with concentrated windings-variation of magnet eddy-current loss due to stator and rotor shapes. IEEE Trans. Ind. Appl. **45**(4), 1334–1342 (2009)
7. As. Masmoudi, Ah. Masmoudi, 3-D analytical model with the end effect dedicated to the prediction of PM eddy-current loss in FSPMMs. IEEE Trans. Magn. **51**(4), 8103711(1-8) (2015)

Printed in the United States
By Bookmasters